Martin Larmont

Medical Adviser and Marriage Guide

Representing all the diseases of the genital organs of the male and female

Martin Larmont

Medical Adviser and Marriage Guide
Representing all the diseases of the genital organs of the male and female

ISBN/EAN: 9783337237004

Printed in Europe, USA, Canada, Australia, Japan

Cover: Foto ©berggeist007 / pixelio.de

More available books at **www.hansebooks.com**

Fiftieth Edition.

PARIS, LONDON, AND NEW YORK.

Medical Adviser and Marriage Guide

REPRESENTING ALL THE DISEASES OF

THE GENITAL ORGANS OF THE MALE AND FEMALE,

THE

MOST COMPLETE AND PRACTICAL WORK ON THE PHYSIOLOGICAL MYSTERIES AND REVELATIONS OF THE MALE AND FEMALE SYSTEMS, WITH THE LATEST EXPERIMENTS AND

DISCOVERIES IN REPRODUCTION.

IT ILLUSTRATES ANATOMICALLY AND FULLY WITH THE PLATES, EVERYTHING PERTAINING TO THE

MALE & FEMALE GENITAL SYSTEMS;

WITH A FULL DESCRIPTION OF

THE CAUSES, SYMPTOMS AND MOST CERTAIN MODE OF CURE, OF ALL THE INFIRMITIES AND DISEASES TO WHICH THEY ARE LIABLE FROM THE SECRET HABITS OF YOUTH, AND EXCESSES OF MATURE AGE,

AS

INVOLUNTARY LOSS OF SEMEN AND IMPOTENCY,

INNOCENT OR UNFORESEEN AFFECTIONS, AND THOSE RESULTING FROM CONTRACTION, AS SYPHILIS, (PRIMARY AND CONSTITUTIONAL,) GONORRHŒA, OR BLENNORHAGIA, (CLAP,) GLEET, STRICTURES, ETC. ETC.

WITH

NUMEROUS CERTIFICATES OF THE MOST UNPARALLELED CURES EVER PERFORMED.

WITH NEARLY ONE HUNDRED ELECTROTYPED ENGRAVINGS.

By M. LARMONT, Physician and Surgeon,
NEW YORK, (FORMERLY FROM PARIS AND LONDON.)

New York:
PUBLISHED BY THE AUTHOR, & DR. E. BANISTER.
1861.

SPECIAL CARD—DON'T FAIL TO READ IT.

M. Ricord, of *Paris*, Acton, of *London*, and the Author, in *New York*, have for many years treated annually a larger number of patients for Genito-urinary diseases, than has fallen to the lot of any other surgeons. The same treatment has been pursued by all; and the frequent recommendation of patients by each to the other, is the best guarantee of the public confidence in our mode of treatment. The Author is happy to state that this treatment is now pursued in all the best European venereal hospitals.

Professional Notice.

The Author and his associate, Dr. E. Banister, devote their time to the cure of all the diseases referred to in this work, at 647 Broadway (up stairs), three blocks above the St. Nicholas Hotel, N. Y. City. We have a number of offices specially appropriated, so that patients never meet each other. For our hours of attendance at the offices, see the New York daily newspapers. *Persons* in need of *medical aid*, can apply to us with the most undoubted assurance of a *cure*, in the most speedy and convenient manner, with perfect privacy.

The afflicted, therefore, who do not wish to submit to the dangerous and injurious treatment of mercury, copaiba, injections, cauterization, quack specifics, antidotes, etc., will read this book or apply to us for a speedy and radical cure.

Patients at a distance, can be *cured* as secretly and as perfectly through the *mail*, and *express*, as by a personal visit—unless surgical operations are required—by giving a full statement of all their symptoms, appearance of the disease, and length of time they have suffered.

N. B.—Our time is so fully occupied for those who send or pay the usual fee upon application, that we cannot respond or notice those who do not pay or remit, by letter or otherwise, the usual consultation fee of five dollars at the time of seeking our advice.

Particular Notice.

In consequence of recent complaints of patients who have lost money letters, and other letters not having been answered by us, I found, on inquiry, that there is an *itinerant Doctor* traveling about the country, and in this city, by the name of Larmont—therefore I wish all letters to be directed to my Associate (who has been with me for several years), Dr. E. Banister, Box 844, Post Office, New York City, as formerly.

M. LARMONT, Physician and Surgeon.

N. B.—Patients are respectfully informed that our offices have been removed from 82 Mercer Street, corner of Spring, to 647 Broadway, *up stairs* (La Farge House block), one block above the Metropolitan, and three blocks above the St. Nicholas Hotels.

PLATE I.

Fig. 1

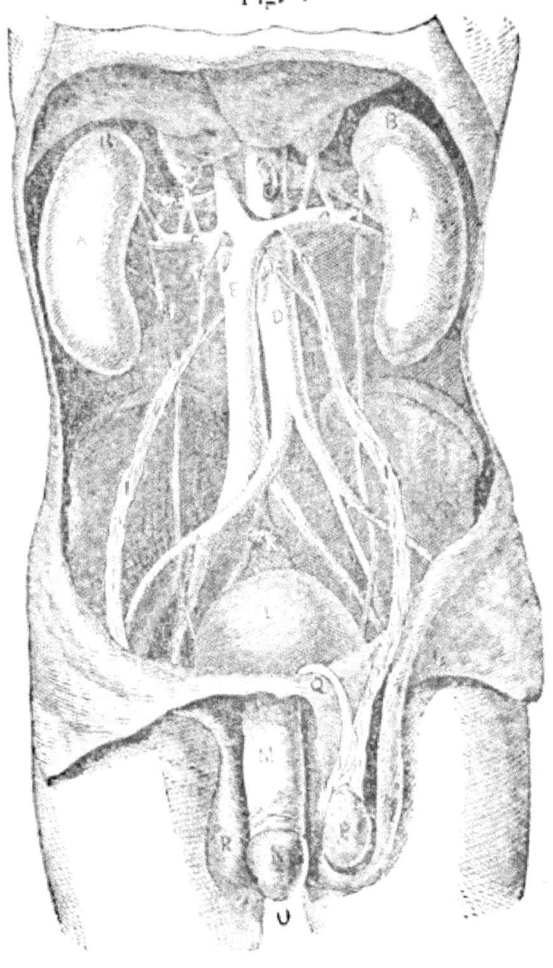

Explanation of Plate I.

FRONTISPIECE.

A Front View of the contents of the Abdomen and Pelvis, and of the Organs of Generation in the Male.

A, A, The kidneys.
B, B, The renal capsules.
C, C, C, C, The renal arteries and veins.
D, The great artery, or aorta.
E, The vena cava.
F, The spermatic artery.
G, G, The spermatic veins
H, H, The ureters.
I, I, The spermatic cords.
K, The rectum.
L, The bladder.
M, The penis.
N, The glans penis.
O, The meatus.
P, The testicle.
Q, The vas deferens
R, The scrotum.

Preface to the 30th Edition.

The rapid sale of the previous editions of the MEDICAL ADVISER AND MARRIAGE GUIDE, has induced the Author to revise and enlarge the work, to render it still more attractive and useful to an intelligent and appreciative public.

In the department of illustrations, many new and beautiful additions have been made, and the entire reärrangement of the plates and figures to give the most convenient and comprehensive study of them by the reader. A larger number of pages have been added than a superficial inspection would seem to indicate, from the fact that vacant pages, and parts of pages, have been filled with interesting cases and other important information.

It is not generally known, that irritation or chronic inflammation of the mucous surfaces of the bladder and kidneys, give rise to severe, painful, and often dangerous and even fatal diseases.

Enlargement and inflammation of the prostate gland, also, when neglected, result in protracted and distressing affections of the genito-urinary apparatus.

These affections of the kidneys, bladder, and prostate gland, give rise to many of the alarming symptoms of stone in the bladder (Gravel), but unlike that melancholy condition, are not only perfectly, but speedily curable by us.

During the past few years, the large increase in the applications for treatment to us of such patients, has enabled us to profit by extensive observation, as is proved by the large number of such patients discharged cured annually.

We wish, in this brief space, to call the attention of the afflicted to the condition of their urine. Healthy urine is a clear, pale amber colored fluid. If, on collecting the first urine voided in the morning in a clean bottle, there is a sediment of white or of a pinkish color, or a cloudy appearance, on settling and becoming cool, it is an evidence of a diseased condition of the genito-urinary system, of a more or less grave character, and should excite prompt attention to the matter by the patient.

In the true diagnosis of these diseases, we are greatly indebted to the microscopic and chemical analysis, without which aid it is often impossible to form an accurate judgment.

PREFACE.

In giving the following pages to the Public, the great aim and object of the Author has been to give a plain, practical treatise, adapted to all circles of society.

All theories, not well sustained by facts and experience, have been discarded.

The most absurd, ridiculous, and mischievous speculations and erroneous assumptions have been published, largely copied and transformed (or rather deformed) from musty antediluvian medical books. It would be better for the public to experience a second deluge, that would sweep away all such trash. It is infinitely more difficult to forget or discard false impressions or erroneous views, than to learn those facts which are correct and of real instruction.

I am led to make these remarks from hearing from all classes, more or less frequently, expressions of wrong views on medical subjects; this must be attributed to the pernicious teachings of foolish books and senseless lectures.

It shall be then, the especial object of this work to

correct such errors, and in so doing our readers will readily see the more plausible and reasonable teachings herein incorporated.

In the first part of this edition great care and pains has been taken to describe the more common derangements of the Male Female and Generative Systems, their anatomy, physiology, and pathological tendencies to disease; and those diseases consequent upon sexual indulgences.

Of Masturbation (Self-Abuse), Emissions, Impotency, etc., etc., the second part of the present edition of this work will fully treat.

Menstruation, and the derangements of the uterine functions have received as ample attention as the character of the work will permit. We have conclusively disproved the falacies and whims which books upon kindred subjects have inculcated: for instance, those directions to prevent conception, which are popularly believed, are not only mischievous and delusive, but a case came under our notice lately, in which the consequences were nearly fatal: This was in a married lady with a deformed pelvis, which precluded the possibility of giving birth to living offspring; yet, deeming herself safe, she followed the directions of a widely disseminated volume, to prevent the mishap, but it had the opposite effect, the accouchment being effected by mutilating the child, and almost immolating the mother. The motive, no doubt, was to gratify the desire that families have to control the number of their offspring.

We have given our experience, and advice founded upon that experience, upon the subject of Conception, and the surest means of accomplishing so desirable an object to thousands of heirless parents.

Sterility, so commonly caused by slight functional derangements in females, is as readily under the control of the competent surgeon as any other functional derangements of other organs, viz., indigestion, torpidity of the liver, bowels, or kidneys, yet from want of analogous reasoning upon causes and effects of uterine affections, medical men often get befogged. and jump at false and injurious conclusions.

Upon the subjects of Medicines, Instruments, or Instructions to produce Abortion, I cannot too severely censure, deprecate and denounce all persons engaged in so abominable and damnable a traffic. Its tendencies are to sap the foundations of all moral and religious society—to ruin the health of females therein engaged—to foster crime—to cloak the iniquities of the wicked—to shield the villainous seducer of the innocent and virtuous, from just odium and public retribution, of the foulest crimes, manslaughter, and even murder!

It is truly lamentable that pious, virtuous, honest, honorable families are destitute of the appreciation of the real magnitude of this monster evil, that the disinclination to have children, induces them to apply to their own family physician to prevail on him to procure abortion.

Having as many as fifty patients constantly under my charge, affected with Venereal and other infectious diseases, the majority of which having been under treatment with other physicians, I have devoted a sufficient space in this work, to guide those who are unfortunate enough to be afflicted with such horrible diseases, to a rapid and permanent cure. The ignorance in the medical profession, of the proper treatment, the awful consequences resulting from the treatment of *Quacks*, and the various humbug cure-all medicines so largely advertised, are truly beyond the comprehension of those who have not the opportunities of knowing its vast extent. As this work will save thousands from such a condition, I cannot but feel that I have fulfilled my duties towards my fellow-beings. My practice is constantly revealing to me the great extent of constitutional disease, arising from neglect or maltreatment of Syphilis. Any physician who will investigate this subject, will be astonished at the mass of human suffering which he can trace to a *Venereal* origin, although its primary symptoms may have been apparently eradicated from the system for many years. Some are indiscreet enough to allow delicacy or shame to prevent them from applying to the proper physician until the poison has acquired such virulence as to justly alarm them; they then frequently apply to some unskillful practitioner, who may temporarily arrest the external symptoms and discharge them cured; matters will thus go on until the malady becomes constitutional, when the patients are at last compelled to place themselves under the treatment of those who, at an earlier period, could have preserved their systems untainted, and their bodies uninjured by the ravages of this most insidious of diseases.

As it is actually necessary, for a thorough understanding of the subject herein treated, I have given, as briefly as possible, a description of the organs of generation, and of their physiology, that any one may clearly understand their importance; furthermore, I have stated the *causes, varieties, symptoms, peculiarities,* and most certain mode of cure.

Every person before entering the Marriage State, should ask themselves if they are in such a state of health that the *Marriage*-bed, will not become a bed o. loathsome disease. I always have some such patients in process of cure, and whose offspring seldom live, but if they do, are the most pitiable evidences of their parents' condition. How degraded and lost to all the finer feelings, which alone can ennoble us, must be that person who will allow, even the remotest doubt to remain unmoved, that they are tainted by disease, or so debilitated by early or guilty excesses, that their progeny will be doomed, but yet dares to offer such a polluted and shattered frame at the pure shrine of love. Such are often the causes of domestic unhappiness, and the horrible diseases which we see children affected with, from generation to generation. Ponder well, then, you who have ever been afflicted with a disease which is sure to become constitutional, if not totally eradicated at the time of infection. Think what your neglect will painfully recall in your own family. That not a single individual need suffer, or have any excuse for their situation, I offer them, THE MEDICAL ADVISER AND MARRIAGE GUIDE, at a merely nominal price - One dollar.

Description of the Plates.

Plate 1—*Figure* 1.—A Front View of the Contents of the Abdomen and Pelvis, and of the Organs of Generation in the Male.—(See minute description opposite the Plate.)

Plate 2—*Figure* 2.—A full and accurate View of the Left Section of the Female Pelvis, in Health.—(See full description opposite the Plate.)

Plate 3—*Figure* 3.—Anatomy and Front View of the Bladder, Penis, etc.

A, The Bladder.—B, The Neck of the Bladder.—CC, The Ureters.—DD, The Vasa-Deferentia.—EE, Seminal Vesicles.—F, Prostate Gland.—G, Urethra.—H, Erector Muscles of the Penis.—II, Cavernous Bodies of the Penis.—KK, Arteries of the Penis.—L, Great Vein of Penis.—MM, Nerves of the Penis.—N, Orifice of the Urethra.—O, The Glans-Penis.—P The Prepuce, or Foreskin.

Plate 4—*Figure* 4.—Anatomy of the Bladder, Prostate Gland, Appendages, etc.

A, The Bladder.—B, Neck of the Bladder.—CC, Ureters.—DD, Vasa-Deferentia.

Plate 4—*Figure* 5.—A Sectional View of the Seminal Vessels and Prostate Gland.

DD, Vasa-Deferentia.—EE, Seminal Vesicles.—FF, Prostate Gland.—G, Urethra.

Plate 5—*Figure* 6.—Penis, Prostate Gland, and Bladder Laid Open.

Letter A, Showing the Mouths of the Seminal Ducts opening into the Urethra.—Letter B, Inflamed Points of the Mucous Surface from Gonorrhœa.—Letter C, Chancre.

Plate 6—*Figure* 7.—REPRESENTS THE VAGINA LAID OPEN BY A LONGITUDINAL SECTION, WITH THE WOMB AND ITS APPENDAGES IN THEIR RELATIVE POSITION.

Plate 6—*Figure* 8.—DIAGRAM OF A SECTION OF THE UNIMPREGNATED GRAAFIAN VESICLE AND ITS CONTENTS, SHOWING THE SITUATION OF OVUM.

Plate 6—*Figure* 9.—INDUSIUM OF THE OVARY, DERIVED FROM THE PERITONEUM.

Plate 6—*Figure* 10.—THE UNIMPREGNATED OVUM SURROUNDED BY ITS PROLIGEROUS DISK, MAGNIFIED ABOUT FIFTEEN DIAMETERS.

Plate 6—*Figure* 11.—A SECTION OF AN UNIMPREGNATED OVUM, REPRESENTING THE THICK EXTERNAL ENVELOPES, CONNECTED WITH THE SURFACE OF THE LATTER.—THE GERMINAL VESICLE.

Plate 6—*Figure* 12.—REPRESENTS WITHIN A SQUARE AREA, THE UNIMPREGNATED HUMAN GERMINAL VESICLE, MAGNIFIED FORTY-FIVE DIAMETERS.

Plate 6—*Figure* 13.—IS A CORPUS LUTEUM, TAKEN FROM A FEMALE WHO DESTROYED HERSELF BY DROWNING, EIGHT DAYS AFTER IMPREGNATION.

Plate 6—*Figure* 14.—REPRESENTS THE INTERIOR OF THE WOMB, WITH THE GERM IN PROGRESS OF FORMATION.

Plate 7—*Figure* 15.—AN OVUM ABOUT EIGHT WEEKS OLD, SHOWING THE PLACENTA WHEN FIRST FORMED.

Plate 7—*Figure* 16.—AN OVUM FIVE MONTHS OF AGE, WITHIN THE WOMB.—(See full description opposite the Plate.)

Plate 8—*Figure* 17.—DISPLAYS A LONGITUDINAL SECTION OF THE UTERUS.

1, Base of the Womb.—2, Mouth of the Womb —3 3, Fallopian Tubes —4, Cavity of the Womb.

Plate 8—*Figure* 18.—ANATOMY OF THE FEMALE PERINIUM

1, 2-6, Sphincter of the Vagina.—3, 4, Erector Muscle of the Clitoris.—5-11, Transverse Muscle of the Perinium.—7, Elevator Muscle of the Anus.—8, Gluteus Muscle.—9, Sphincter of the Anus.—10, Junction of the Sphincter of the Anus, and the Sphincter of the Vagina.—12, Adductor Muscle.—13, The Gracilis Muscle.

Plate 9—*Figure* 19.—REPRESENTS THE FŒTUS AT MATURITY, AS SEEN FOLDED UP AND IN ITS NATURAL POSITION, IN THE WOMB.

Plate 10—*Figure* 20.—INFLAMMATORY STAGE OF GONORRHŒA CAUSING PHYMOSIS.

Plate 10—*Figure* 21.—GONORRHŒA WITH CHORDEE.
" " 22.—GONORRHŒAL OPHTHALMIA.
" " 23.—GONORRHŒAL DISCHARGE MAGNIFIED.

Plate 11—*Figure* 24.—EXTERNAL GENITAL ORGANS OF THE FEMALE, REPRESENTING PRIMARY SYPHILIS AND GONORRHŒA AT THE SAME TIME.

Plate 12—*Figure* 25.—STRICTURES IN THE URETHRA.
" 13— " 26.—STRICTURE IN THE URETHRA, AND ENLARGEMENT OF THE PROSTATE GLAND.

Plate 14—*Figure* 27.—SWELLED TESTICLE FROM URETHRITIS.

Plate 14—*Figure* 28.—HYDROCELE, OR DROPSY OF THE TESTICLE.

A, Penis drawn almost into the Body.—B, Scrotum distended to its utmost extent by the Water.—C, The Testicle.—D, The Vas-Deferens.

Plate 14—*Figure* 29.—EXCRESCENCES — VEGETATIONS, OR WARTS ON THE PENIS.

Plate 15—*Figure* 30.—PHYMOSIS.
" " 31.—DEEP-SEATED SYPHILITIC ULCERS ON THE SCROTUM.

Plate 15—*Figure* 32.—CLAP, CHANCRES (SYPHILIS), AND PARA-PHYMOSIS IN SAME PATIENT.

Plate 15—*Figure* 33.—Chancres (Pox) causing Phymosis.

Plate 16—*Figure* 34.—Simple Chancre (Pox) six hours after Inoculation.

Plate 16—*Figure* 35.—Simple Chancre (Pox) thirty hours after Inoculation.

Plate 16—*Figure* 36.—Simple Chancre (Pox) two days and six hours after Inoculation.

Plate 16—*Figure* 37.—Simple Chancre (Pox) three days and six hours after Inoculation.

Plate 16—*Figure* 38.—Simple Chancre (Pox) four days and six hours after Inoculation.

Plate 16—*Figure* 39.—Simple Chancre (Pox) five days and six hours after Inoculation.

Plate 16—*Figure* 40.—Simple Chancre (Pox) six days and six hours after Inoculation.

Plate 16—*Figure* 41.—Simple Chancre (Pox) ten days and six hours after Inoculation.

Plate 17—*Figure* 42.—Chancres (Pox) in the Urethra and Bladder.

Plate 17—*Figure* 43.—Syphilitic Ulcers on the Tongue and in the Throat.

Plate 17—*Figure* 44.—Syphilitic Discharge magnified
" 18— " 45.—Destruction of the Eye by Syphilis.
" 18— " 46.—Constitutional Syphilitical affection of the Eye.

Plate 18—*Figure* 47.—Syphilitic Hydro-Carcocele.
" 19— " 48.—Syphilitic Pustules previous to the nose being destroyed.

Plate 19—*Figure* 49.—Nodes on the Forehead.
" 19— " 50.—Destruction of the Nose by Syphilis.

Plate 19—*Figure* 51.—CARIES IN THE TEETH FROM SYPHILIS.
" 20— " 52.—SPERMATORRHOEAL OPHTHALMIA IN CONSEQUENCE OF MASTURBATION.

Plate 20—*Figure* 53.—HORRIBLE APPEARANCE OF THE FEATURES CAUSED BY SELF-ABUSE.

Plate 21—*Figure* 54.—VARICOCELE PRODUCED BY SELF-POLUTION AND INVOLUNTARY LOSS OF SEMEN.

Plate 21—*Figure* 55.—TOTAL RELAXATION OF BOTH TESTICLES FROM ONANISM AND SEMINAL DEBILITY.

Plate 22—*Figure* 56.—MICROSCOPIAL VIEW OF SEMEN AND SPERMATOZOA.

Plate 22—*Figure* 57.—THE STRUCTURE OF THE TESTICLE INJECTED WITH MERCURY, AND ITS SEVERAL PARTS UNRAVELED.

1, 2 2, Tubuli Seminiferi.—3, Vasa-Recta, forming the Rete-Testis.—4, Corpus-Highmorianum.—5, Vasa-Efferentia, forming the Coni-Vasculosi.—6, A single tube formed by the junction of the Vasa-Efferentia. This tube then becomes convoluted upon itself to form the Epididymis.—7, 8, Beginning of the Vas-Deferens.—9, The Vas-Deferens becoming a straight, isolated tube in its ascent to the Abdominal Ring.—10, Spermatic Artery.—11, Spermatic Cord, dissected and spread out.

Plate 22—*Figures* 58, 59, 60, 61, 62, 63, 64, and 65.—SPERMATOZOA AND GRANULA SEMINIS.

Plate 22.—*Figures* 66, 67, and 68.—REPRODUCTION CELL, CONTAINING THREE GRANULA SEMINIS.

Plate 23—*Figure* 69.—SPERMATOZOA FROM THE HUMAN TESTICLE.

Plate 22—*Figure* 70.—FASCICULUS OF HUMAN SPERMATOZOA MAGNIFIED ABOUT A THOUSAND TIMES.

Plate 23—*Figure* 71.—REPRESENTS THE DIGESTIVE TUBE FROM THE ESOPHAGUS TO THE ANUS.

Plate 24—*Figure* 72.—THE TESTIS IN HEALTH.

PART I.

PARTICULAR CAUTION.

Avoid all quack or advertised remedies for any disease, especially venereal, and all of the instruments advertised to cure Seminal Emissions, as they are injurious quack humbugs.

Within the past year or two, there have arisen a number of mushroom doctors, pretending to be from Hospitals in the different cities of Europe, or students of eminent men, such as Ricord, Acton, etc., etc., using their medicines, medicated bougies, rectum suppositories, etc., etc., whose injured and uncured patients we are curing every week. Another caution, to save yourselves from imposition in our own building, be certain you see our No., 647 Broadway, and the silver plates on each side of the door, stating the "Physician's Consulting Rooms are on the second floor," where, upon ascending the stairs, as directed, you will see our names upon silver door plates, on the doors of the offices.

PLATE II.

Fig. 2

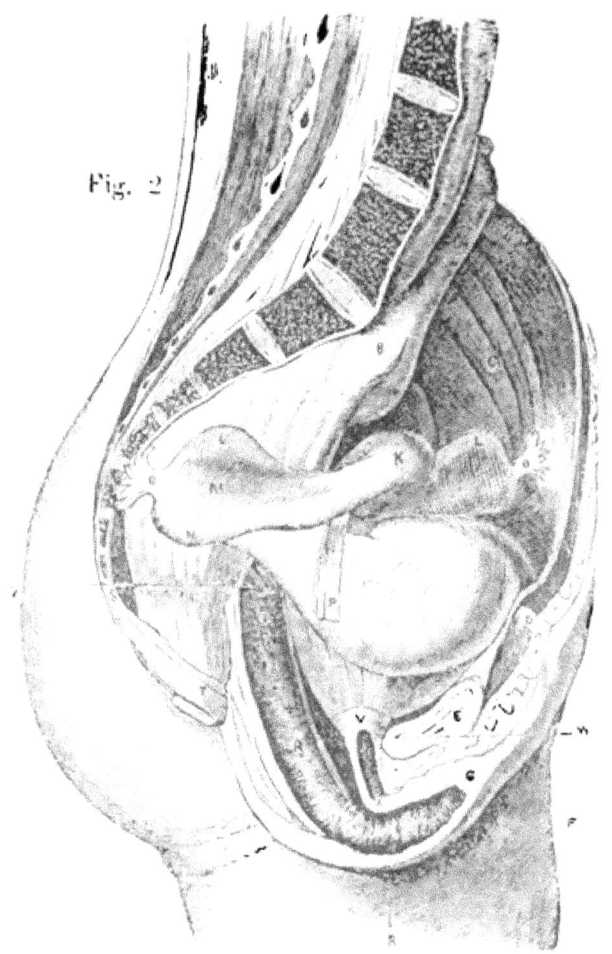

Explanation of Plate 2.

A full and accurate View of the Left Section of the Female Pelvis, in health.

A, The fourth lumbar vertebra.

B, The rectum.

C, The left iliac fossa.

D, The rectus abdominus muscle, arising from E, The symphysis pubis.

F, The mons veneris. G, The clitoris.

H, The left nympha. I, The labia externum.

K, The fundus uteni.

L, L, The ovaries brought upwards.

M, The posterior surface of the right broad ligament.

N, The right Fallopian tube turned downwards.

O, O, The fimbriated extremities of the tubes.

P, The right round ligament.

The dotted line crosses the fundus of the bladder

Q, The os uteri.

R, The vagina.

S, The point of the coxcyx.

T, The sphincter ani.

V, The sphincter of the bladder.

W, The urethra—the dotted line crosses the perineum.

X, The maetus urinarius.

Introduction to the 30th Edition.

In no class of diseases are the functional derangements of the stomach, bowels, kidneys, and bladder, so troublesome and annoying to both patients and physicans, as in seminal affectons, and probably from sympathy extending through the entire *nervous system.*

It is a frequent occurrence, to have calls for relief, of frequent and painful micturation, or irritable bladder, the sufferer not having suspected that aught else ailed him ; but this is one of the commonly occurring complications of disease previously existing in the genito-urinary organs, either from spermatorrhœa, gleet, or stricture, having caused inflammation of the mucous membrane, which is so liable to extend to adjacent parts, for it is the same continuous tissue which lines the whole system of these emunctory organs.

Excesses in eating or drinking, or of excessive venery, in these patients, excites inflammatory action even in these and more distant organs.

Such persons, are those who often contract leucorrhœa from their wives, and simply from the existing predisposition to inflammation of the uerthral mucous membrane. When gonorrhœa does not promptly yield to treatment, we know that there is something in the back-ground that will soon *loom up* in total disregard of the laws of perspective ; delaying and tediously protracting the cure, unless *two birds* are killed at one *shot.*

THE

ANATOMY OF THE GENERATIVE ORGANS

OF THE

MALE AND FEMALE.

NOT only have the ancient and modern writers admitted the importance of the sexual organs being preserved in a perfect state of health and vigor, but every person possessed with a mind capable of a moment's reflection will see its importance on mankind at the present time, and particularly for the future. The proper performance of the special functions with which they are charged has ever been considered essentially necessary to the health and well-being of the economy, both physical and mental. They are parts of admirable construction, form, and use; and constitute a striking evidence of the wonderful skill and contrivance in the adaptation of a special mechanism in the system for the performance of one of its most important and essential functions:—that of the propagation of the species. Unequaled in the delicacy of their texture, and the comparative minuteness of their structure, their peculiar fitness for the functions assigned them in the economy, when they are in a state of perfect integrity, excites the astonishment and admiration alike of the anatomist and the philosopher. Their very complexity, while it renders them liable to many disorders, by any of which their utility may be impaired, is wisely rendered subservient to the important purpose of separating and purifying the vivifying fluid.

Like that complex and delicate piece of machinery —a watch—constructed by human skill, the organs of generation—a still more complex and more delicate apparatus, created by the Divine will—are liable to derangement and impairment of functions and structure from many causes, the nature and effects of which will be investigated in the following pages. In order, however, that these may be fully and clearly understood, it will be advisable to preface the observations we propose hereafter to offer respecting them, by some notice of the anatomical arrangement and physiological action of the organs which are immediately subservient to the function of generation, and also of those which are only indirectly connected therewith.

The parts in man which are immediately connected with the functions just alluded to, are the *testicles*, by which the *semen* or *seed* is secreted, and of their appendages, through which the seminal fluid is transmitted to the *urethra* at its origin near the *neck* of the *bladder*, and of the *penis*, by means of which the act of copulation takes place, and through a *canal* in the under part of which, called the *urethra*, the seed is conveyed from the receptacles in which it is retained, to those organs, in the female, which are engaged in the functions of generation.

The urinary organs, both male and female, may be regarded as subsidiary to this function, and many of the diseases to which they are liable, exert an influence on its performance, and not unfrequently produce impotence, either temporary or permanent, according to the nature, (treatment, if employed,) and severity of the disease.

The KIDNEYS, which are the organs solely engaged in the secretion of the urine, are glandular bodies of an oblong shape, seated on either side of the spine, upon and below the two last ribs, and behind the stomach and intestines; the right kidney is also under the liver, when the man is in an erect position, and the left under the spleen: the right kidney is generally the lower and the larger. It is said that these organs are considerable in size in those persons whose passions are very strong, and almost uncontrollable, than they are in those who are less addicted to sensuality. In shape the kidney resembles the kidney-bean; its structure is almost wholly made up of arteries and veins, with a few small branches of nerves, derived partly from those which are connected with the ribs, and thence called intercostal, and partly form a branch from the stomach, thus causing a great sympathy between those organs. The artery by which the kidneys are supplied with blood, which is partly used for the support of the organ, and partly for the secretion of urine, is derived directly from the aorta, or great artery of the body. When it enters the kidney, which it does about its middle, it divides into branches which again are divided into smaller ones, and these into smaller still, until they terminate in vessels so exceedingly minute as to be invisible to the naked eye. From these the veins are formed, and by these the urine is secreted, and falls by drops into a pouch which is situated about the middle or lower part of the organ, and which forms the commencement of the ureter. The vein joins the great cava vein, and discharges its blood into what is called by anatomists the great portal system, by which it is conveyed to the liver, after it has been freed in the kidney from a certain portion of its serum, and also from certain salts. The nerves of the kidneys are few and small.

The URETERS are long hollow tubes, and constitute the continuation of the pelvis of the kidneys. There is one on each side of the body, and they pass downwards, and slightly inwards to the back and lower part of the bladder, which they pierce, running between its coats for about an inch, so that if the bladder should become exceedingly distended, its contents would not be forced back into these tubes. They are well supplied with branches of arteries, veins and nerves, and their sensibility, in a state of disease, is considerable. Their use is to convey the urine from the kidney into the bladder.

The BLADDER is situated in that part of the body called the pelvis. It is of considerable size, and admits, in some instances, of distension to a degree that would hardly be credited, were it not a well known fact. Such distension is at the risk of health and life. This organ in man lies directly on the bowels. It is of an oval shape, constitutes the great receptacle of the urine. The bladder is well supplied with arteries, veins and nerves. It has three coats, one of them being composed of muscular fibres; its construction causes the expulsion of the urine; it has on that account been called the *detrusor urinæ*.

The neck of the bladder—which in man is longer and narrower, and in woman is shorter and wider—is surrounded by a sphincter muscle, by which the

PLATE 3.

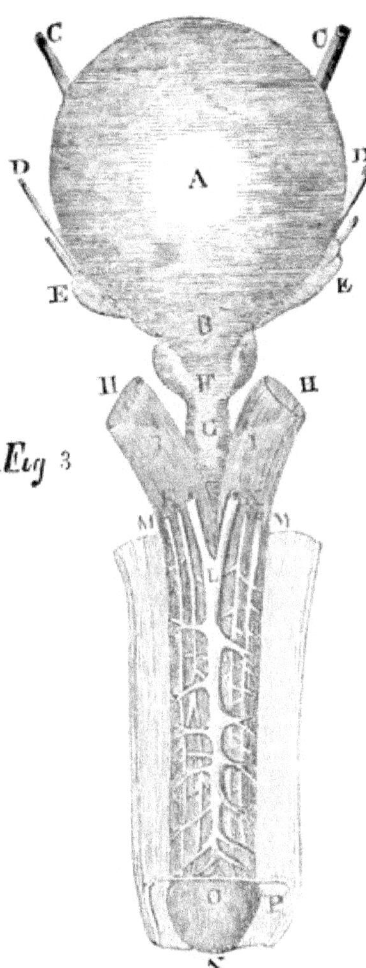

continued running away of the urine is prevented, unless from disease the muscle has become useless.

The process by which the secretion of the urine is effected, is one of exceeding interest. The blood from which it is to be separated, is conveyed to the organ by the venal artery, which divides into branches, supplying different parts of the organ; and these again in their turn form arches of communication with each other, whence spring minute arteries or branchlets; these again constituting a complete net-work of vessels by a general inosculation. They terminate in the commencement of veins, and also in uriniferous tubes, by which latter the separation of the urine is effected. The crypts or *cryptæ*, small round or oval bodies, which are found everywhere in the net-work of vessels just spoken of, and which consist almost solely of vessels, are by some supposed to be the origin of the uriniferous tubes. The tubes terminate in a mammillar process, which projects into a small membranous bag, called from its shape the infundibulum or funnel; into this bag the urine passes from the uriniferous tubes; it is thence conveyed to the larger pouch called the pelvis, and afterwards through the ureter into the bladder. The last named pouch, like the pelvis of the kidney, the ureters, bladder and urethra, is defended from the acrimony of the urine by a secretion of mucus which lines and sheathes its inner coat.

In patients laboring under some difficulty from stricture in passing urine, the mind will often greatly increase the secretion of that fluid, and multiply the calls to pass it from the body. This will be exemplified in the chapter on strictures.

The SCROTUM, is a bag of skin, divided about the

middle by a septum, so as to form two cavities, in each of which a testicle is contained. The situation of this septum is marked externally by an irregular line called the raphae. The contraction or corrugation of the scrotum, which occurs at times, is said by some anatomists to depend on the action of a muscle which they call *dartos*. The testicles or organs which secrete the semen, are nourished and supplied with blood by long and tenaceous vessels which arise from the main arterial trunk, and are called spermatic arteries; the blood which they thus receive, serves for elimination and secretion of the seed, a process which is effected by the peculiar action of the testicles, and which secreting power affixes to these organs a value and importance in the human frame, not even second to that which attaches to those generally regarded by anatomists as the most noble, being those, the destruction or serious impairment of the functions of which may involve loss of life.

The ancient Romans would not allow any one to bear witness against another in a court of justice, unless he were perfect in the organs of generation—unless the testicles were sound and entire. The papal clergy so far carry this rule into effect, that no one can be admitted a member of their priesthood, against whom a similar defect can with truth be alleged.

Occasionally the testicles, which before birth are lodged within the cavity of the abdomen, do not descend into the scrotum or purse, but remain in the belly, generally within what is called the abdominal canal. Sometimes one only is retained in the abdom n, and that generally the left. In this situation they are exposed to various causes of disease, and although not absolutely deprived of the power of secreting seed, yet their action is generally more or less imperfect, in all probability from the compression they undergo, and the constant irritation to which they are subject, from the narrowness of the canal by which they are, in fact, somewhat elongated, and flattened, and smaller than usual.

The spermatic artery is a long undulating, and tortuous vessel. The blood which is thus conveyed to the organs, after having been employed by the testicles for the separation and secretion of the seed, is re-conveyed in a refuse state by other vessels, called the spermatic veins, back to the general circulatory system in the body. The double set of vessels, the veins and arteries, the old anatomists call the *vasa deferentia*, as being the parts principally concerned with the testicles in the preparation of the seed.

The spermatic arteries and veins are remarkable for their smallness, which prevents their containing more than a small quantity of blood at a time. They pass obliquely downwards and outwards, behind the peritoneum, and are contained in a common protecting sheath with the veins, forming with the nerves of the testicles the spermatic chord; they then run over the psoas muscles and ureters, and pass out through the rings of the abdomen and abdominal canal, over the os pubis, and into the scrotum, and supplies the testes, which also receives blood from the artery which supplies the vas deferens. The latter named organ, which is invested in its own sheath, called *tunica vaginalis*, is composed of the body of the testicle, and the epididymis, the latter being situated at the upper part. It consists of an infinite number of small tubes (seminiferous) which terminate in the epididymis. These tubes are convoluted on each other, and closely connected together, but which, unraveled and injected with quicksilver, will extend to a considerable length.

The spermatic veins arise in three sets from the testicle, two of which soon unite. They are exceedingly tortuous in their course, and fully anastomose with each other, while in the lower part of the cord, but these inter-communications cease after they have entered the abdominal canal, on leaving which, while crossing the psoas muscle, they unite together and form one vein, which, on the right side, terminates in the lower vena cava, and on the left in the vein which arises from the kidney on that side. The larger veins are provided with valves. The nerves of the testicles are principally derived from those which supply the kidneys. They take the same course as the spermatic arteries, and constitute with them and the veins the spermatic cord. The spermatic nerves are finally distributed to the substance of the organ, to the due performance of the function of which they are subsidiary.

According to the correct laws of nature, each male person should have two testicles, one in each side of the scrotum, but I have had patients with but one, and some with three. It has been asserted that some males have been known to have four or five, but I doubt its truth. When a man has but one testicle, it is generally larger than is natural; and in those I have seen who had three, one was generally smaller than the other two, but, in one case, one was larger than either of the other two. Such person's passions are generally stronger than those who have two; yet those who had but one, differed but little from the major part of those who have two. Instances have been known of males never having any testicles from birth.

Where these most important organs are natural in size, number and general appearance, they are generally nearly two inches in length, one and a half in the transverse direction, and one in thickness. The tunica vaginalis, or investing membrane of the testicles, consists of two layers, the inner one directly enveloping the testicle. It secretes a kind of serum which serves to lubricate it. Between the two layers of the vaginal tunic, is contained the fluid hydrocele of the scrotum. In some cases the cavity formed between the two layers of this membrane, remains continuous with the cavity of the abdomen. In such instances, there is the double danger of the occurrence of what is called congenital rupture. (Hernia.)

Between the testicle and the tunica vaginalis, there is another tunic or coat, called the *tunica albuginea*, which is smooth, white, and inelastic. It completely covers the testicle, but not the epididymis. The testicle is also invested and protected by a muscle, called the *cremaster*. It expands all around the tunica vaginalis, which it closely embraces, forming a hollow muscle, within which the testicle and its tunics are contained, and which, when it is in action, contracts and draws the organ it encloses upwards to the abdomen, sustaining and compressing it, and forcing out along the *vas deferens* the semen previously secreted by the organ. The *cremaster muscle* is small and indistinct prior to puberty; after that period it is greatly developed in persons who are very muscular.

It has already been observed, that the substance

of the testicle consists of an infinite number of small tubes, which are called the *tubule seminiferi*. These are very numerons; the number has been calculated at eight hundred and forty, and their entire length at one thousand seven hundred and fifty feet, the mean length of each duct being twenty-five inches. They communicate readily with each other, and thus constitute one vast net-work of communication. Their size is greater in an active adult in the prime of life, while the organs are in full vigor. They differ in the testicles of the same individual. Two or more tubes unite and form a conical lobe, of these, there are between four and five hundred in each testicle.

The Epididymis, which it has been stated is seated at the upper and back part of the testicle, is the continuation of the numerous seed-bearing tubes; it descends along the back part of the testicle, gradually becomes larger in diameter, but less convoluted until it begins to ascend, when it obtains the name of vas deferens. It is no longer than the testicle. It consists principally of seminal canals, from which arise in the after part of the right testes, the *vasa efferentia*, or defferent vessels, of which tubes there are generally twelve. Their average united length, is nearly eight feet, the separate length of each, being rather more than seven inches.

The *vas deferens*, the excretory duct of the testicle, forms a constituent part of the spermatic cord, and is readily distinguished from the arteries, veins, nerves and absorbents, by its cartilaginous feel. It terminates in the seminal vesicle, immediately above and behind the prostate gland, and with it, forms the ejaculatory canal, which perforates the prostatic part of the urethra.

The testicles in the fœtus, are situated in the abdomen, immediately below the kidneys. The epididymis is about one-third larger relatively to the body of the testicle, than it is in the adult. Towards the close of the period of puberty, the testicles are generally found in the scrotum. The non-descent of both testicles is of comparatively rare occurrence. One sometimes remains permanently fixed in the situation which it occupied when the child was born, but it occasionally descends prior to puberty, most generally between the second and the tenth year. An operation for the descent of a testicle, is tedious, severe, and sometimes fatal The testicles not having descended into the scrotum, are not generally deprived of their power of generation. In some cases, their integrity is full. It occasionally happens that the testicles do not attain their full size and power of secreting semen. Such instances are not beyond the influence of the proper treatment, however, unless they occur in the persons of idiots. This is treated upon in the subsequent pages of this work.

Semen will not combine with water at any temperature, from zero to the boiling point, unless it has previously been liquified in nitric or sulphuric acids. The amount of seminal fluid emitted during the act of sexual congress, varies from one to two or three drachms. As stated in the following pages of this work, healthy semen contains animalculæ.

The spermatozoa are imperfect or deficient in the semen of mules or hybrid animals. Hence depends, in all probability, the impotence or sterility of those creatures. They are generally utterly incapable of generation There are, however, instances both among the mammalia and birds, of individuals belonging to species universally held to be distinct, uniting and producing young, which again were prolific. The mule can engender with the mare, and the she mule can conceive. They occur, however, more frequently in warm countries. Buffon says, the offspring of the he goat and ewe, possess perfect powers of re-production. We might expect these animals, with the addition of the Chamois, to copulate together easily, because they are nearly of the same size, very similar in internal structure, and accustomed to artificial domestic life, and to the society of each other from birth upwards. There is a similar facility in some birds, when such unions are often fruitful, and produce prolific offspring. The cock and hen canary birds, produce with the hen and cock, siskins and goldfinch; the hen canary produces with the cock chaffinch, bullfinch, yellow-hammer, and sparrow. The progeny in all these cases is prolific, and breeds not only with both the species from which they spring, but likewise with each other. The common cock, and the hen partridge, as well as the cock and Guinea hen, and the pheasant and the hen, can produce together.

Notwithstanding all these, and perhaps other examples which might be adduced, the general rule is, that hybrids are incompetent to perform the act of generation, so as to produce offspring; and it is a wise provision of nature that such should be the case, to prevent the world being inhabited by monstrous creatures, as would be the case, were it the general rule that fecundation followed the act of copulation, when practised by the offspring of parents of different species.

The VESICULÆ SEMINALES, or seminal vesicles, are two sacs or oblique bags, behind and below the bladder, between it and the rectum, and closely connected with it by cellular tissue. That part which is applied against the bladder is concave, the opposite surface convex. They occupy an oblique position, their lower extremities being separated only by the different vessels, while their upper ends are at a considerable distance from each other. The latter are the larger, and their greatest breadth is generally three or four times less than their length, and their thickness is about one-third of their breadth; they are about three fingers, breadth in length; the contents pass through vesicles from one part of the tube to the other.

The seminal vesicles have two coats; on the surface of the inner one small cells exist. It, no doubt, is a secreting membrane. The seminal vesicles are well supplied with arteries, veins, nerves and absorbents. Near the prostate cells cease to appear; the vesicle contracts, and forms a kind of duct which unites with the vas deferens at a very acute angle, the place of union being marked by a projecting septum or valve, by which the contents of the defferent vessels are directed into the seminal vesicle.

The ejaculatory duct thus formed by the union of the vas deferens and seminal vesicle, is from half an inch to three quarters long; it continues to become narrower as it passes behind the third lobe of the prostate, perforates that body, and running some way along the under surface of the urethra, enters that canal obliquely by a small opening on the side of the caput gallinaginis. The junction of the two vessels which form this common duct, is such, notwithstanding the acuteness of the angle, air gently thrown into the vas deferens by a blow pipe, will inflate the seminal vesicle before it enters the urethra, but if thrown with violence, it will immediately inflate both the urethra and seminal vesicle. The seminal vesicles are very large in the boar, and divided into cells of considerable extent, having one common duct. They have no communication with the vas deferens in the rat, nor in the beaver. In the Guinea pig, they constitute long cylindrical tubes, and have not any communication with the defferent vessels. These facts do not afford conclusive proof, however, that the seed may not pass into the vesicles from the defferent vessels in the human subject. Notwithstanding the acuteness of the angle between the two vessels at their junction, from the length of the common tube the wideness of that part of it formed by the vesicle wher the two vessels meet, and the very small aperture by which it opens into the urethra, the fluid, which form the length and contortion of the seminal tubes, must pass very slowly from the testicles, will insinuate itself much more readily through the large communication with the vesicle, than through the very small ones with the urethra, unless it be prevented from so doing by the vesicle attempting to throw its contents into the urethra at the same time. During coition, this attempt is made, and both fluids pass at once into the urethra, where the fluid secreted by the vesicles being added to that coming from the testicles by the defferent vessels, between them, a proper quantity is produced to distend sufficiently the sinus of the urethra, that the muscles of ejection may act on its contents with more power.

From the frequent excitement of the passions, and their gratification being denied in the civilized state of human society, fluid must often be secreted in the testicles at times, when it cannot be naturally evacuated; and although the accumulation of it in this organ sometimes produces tension and pain, the fullness of the vessels often subsides without these unpleasant symptoms having taken place. Thus, when the vis a tergo no longer drives the semen slowly on, the muscular properties of the vas deferens may assist in conveying that fluid on towards the vesicles, which may receive it until the time of ejectment arrives. They may thus under particular circumstances,—more likely to occur in the human species than in brutes,—be employed as reservoirs, although their ordinary use may be to secrete a fluid which, mixing with the semen during coition, may render the act more perfect, and more likely, therefore, to produce fecundation.

An additional reason may be adduced in support of the theory, that the seminal vesicles act as reservoirs for the seed in man, in the well known fact, that animals possessing a penis, but destitute of seminal vesicles, remain for a long time in sexual contact, because the fluid necessary for fecundation, from the long course it has to take during copulation, only flows from the urethra drop by drop.

A distinct communication between the seminal vesicles and the deferent, takes place only in man, and in those animals which most resemble him in form, as in the whole tribe of the simiae. The vesicles are altogether absent in the lion, panther, cat and dog.

The prostate gland, in shape and size, resembles a chestnut. It is situated below and behind the bladder, and above and in front of the rectum. The base inclines upwards and backwards, the apex pointing downwards and forwards. A notch in the middle of the base, divides the prostate into two lateral lobes, immediately above which are the lowest parts of the deferent vessels, and seminal vesicles, the ducts of which begin to perforate the gland in the middle of the notch, and then pass into the under part of the urethra, where it is surrounded by the substance of the gland. The neck of the bladder is surrounded by the prostate, as is also the commencement of the urethra. When the prostate gland becomes enlarged from diseases, it passes upward towards the cavity of the bladder, immediately behind the commencement of the urethra, and occasionally bends over that opening, acting as a sort of valve to prevent the expulsion of the urine. The veins and absorbents of the prostate are numerous, and empty themselves into those which are connected with the bladder. The fluid which it secretes is of a white or cream color; it is viscid, and has a slightly salt taste. Its use seems to be to lubricate the surface of the urethra, along which the semen is to pass. It is thrown out in considerable quantity, when the parts are in a state fit for immediate copulation; much of it then unites with the seminal fluid, and is discharged with that fluid when emission takes place.

The fluid of the prostate, like that of the seminal vesicles, is not absolutely necessary for the purposes of generation in all animals which possess testicles; and although the gland is found in man and the tribes of the simiae, the lion, dog, etc., it is not present in the bull, the buck, and ram, and goat, and most probably in all ruminating animals

PLATE 4.

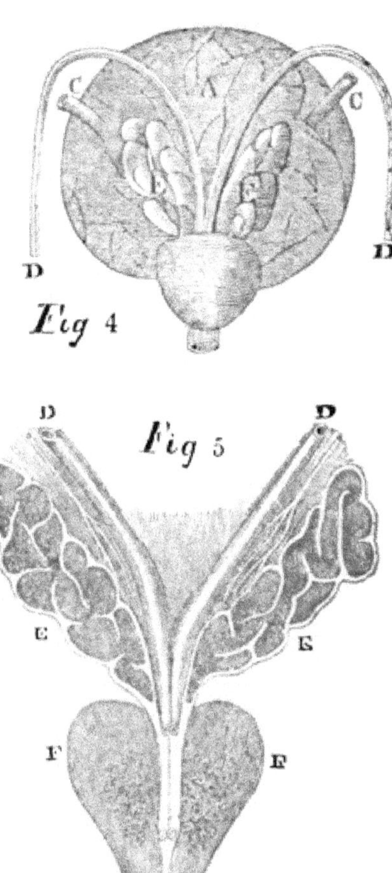

Both the gland and the seminal vesicles are wanting in birds and amphibious animals, and in fish which have testicles, as the ray kind. The prostate is said to be double in the elephant, camel, horse, and some other animals. The semen is never evacuated, but where the liquor of the prostate gland goes before, and follows after it. It is obvivious, therefore, how powerfully it must conduce to health, to have the secretion of this gland in a sound and pure state, as it is so intimately connected with the finest functions in the animal economy. The seed and secretion of the prostate gland are intimately mixed together in the urethra. The appearance of the secretion of the prostate, when diseased, nearly resembles putrid matter. It is plentifully secreted in good health, and seems intended by nature to be a vehicle to dilute, nourish, and convey the thick and ash-colored concocted semen. We have seen in the most healthy men, who have long abstained from venery, a running of the humor from this gland, from its being in a relaxed state, during which the semen will be emited by the slightest straining, and from ideas of the mind, both while awake and asleep. The sooner the patient gets this relaxed state restored the better. I very often cure patients who had been under the charge of physicians, surgeons, and professors of our colleges, and who treated them for a venereal or gonorrhœal affection, when, in fact, it was a diseased prostate and the neighboring parts. Errors of this kind produce great injury. Eunuchs often eject prostatic liquor when they have an erection; geldings often do the same when they strive to leap. Good semen cannot be found when these parts are diseased; great caution should, therefore, be observed by all those entering the marriage state, to be well assured that this humor of the prostate is in a sound and healthy state; various evils will arise in consequence, especially sterility and impotence.

Healthy men continually separate semen from the blood, which, being retained and inspissated, like the white of an egg or starch, would be most immoveable, if it were not for the more thin juice of the prostate gland, when in a sound state, which mixes with it, and serves to lubricate the urethra, almost like an oil. Besides this, as the animalculæ must stay a long time, perhaps, before it arrives in the uterus or womb, it seems necessary for it to be provided with a suitable aliment; for, unless nature nourished the animalculæ when formed, it would certainly perish or become extinct: and this nutritious liquor is that of the prostate gland, which in some animals is larger than are the testicles themselves.

Cowper's glands, which are situate between the bulb of the urethra and the membranous portion, are about the size of two small garden peas. They open into the canal by two small ducts, and appear to secrete a mucus which serves to lubricate the urethra.

The urethra, a membranous canal, extending from the neck of the bladder to the end of the penis, is divided into the prostatic, membranous, bulbous, and pendulous portions. Its coats are the same as those of the bladder, of which it is apparently a prolongation. The first or prostatic portion, commencing immediately from the neck of the bladder, is surrounded by the prostate. On the under side of its internal surface there is a projecting body, on the sides of which the common ducts of the deferent vessels and seminal vesicles open into the canal, as also the ducts of the prostate.

The portion of the urethra between the prostatic and bulbous portions, is called the membranous; and the reason that has been alleged for this is, because its circumference is less than that of any other part of the canal. Its length is generally about an inch when the penis is in a state of erection; when otherwise, it is somewhat less. It is cylindrical in form for about half its length. The urethra soon after takes the name bulbous, when it meets with the pendulous portion of the bulb, the substance of which, however, it does not enter until it reaches the arch of the pubis. At this part it is attached to the symphysis by muscular fibres. These muscles are influential in the expulsion of the semen. The urethra at this part enlarges somewhat at its under part, forming a kind of sinus. The canal afterwards bends forwards, and is surrounded by the spongy bodies through its course along the under surface of the penis.

The whole of the internal surface of urethra is abundantly supplied with mucus to defend it from the acrimony of the urine. It is secreted partly by vessels which form small projections on the inner surface of the canal, as shown in the engraving, and partly by glandular structures situated at the bottom and sides of the very numerous lucunæ, or depressions, dispersed over every part of the internal membrane, the openings of which are directed towards the termination of the urethra, so that the mucus is pressed out of their cavities by the urine as it flows from the bladder.

The urethra is very vascular, and possesses a certain

degree of elasticity. Its membranes are very thin, and almost transparent, and without fibres, so that in itself it does not possess the power of muscular contraction and relaxation. It is, however, provided with muscles, the action of which is to assist the expulsion of the urine, and also of the semen during copulation. The whole length of the penis is about twelve inches, though it varies much in different individuals.

The PENIS consists of the cavernous bodies, (Corporo cavernosa,) and of the spongy body, (Corpus spongiosum,) the latter terminating in the gland or glands. The cavernous bodies constitute the upper part of the penis; in the upper groove there being a large vein, two arteries, nerves and absorbents, and in the lower the spongy body surrounding the urethra. The convex conical surface of the gland is covered by a fine membrane, in color resembling the red part of the lips. At its base, or corners, there are rows of projecting papillar, which secrete a sebaceous matter, having a peculiar smell. The gland, which possesses exquisite sensibility, is protected by the loose covering called the prepuce or foreskin, which is tied to the penis, immediately below the orifice of the urethra, by the band called frænum: this limits the motion of the prepuce, and tends to keep it in its proper place.

The spongy substance of the urethra, which forms the *glans penis*, is covered externally with an exceeding thin membrane or cuticle, under which are placed the very sensible nervous papillæ, which are the chief seat and cause of pleasure and pain in this part. We may now understand why many, in the venereal act, have not the glans distended, though the whole penis is at the same time turgid,

PLATE 5.

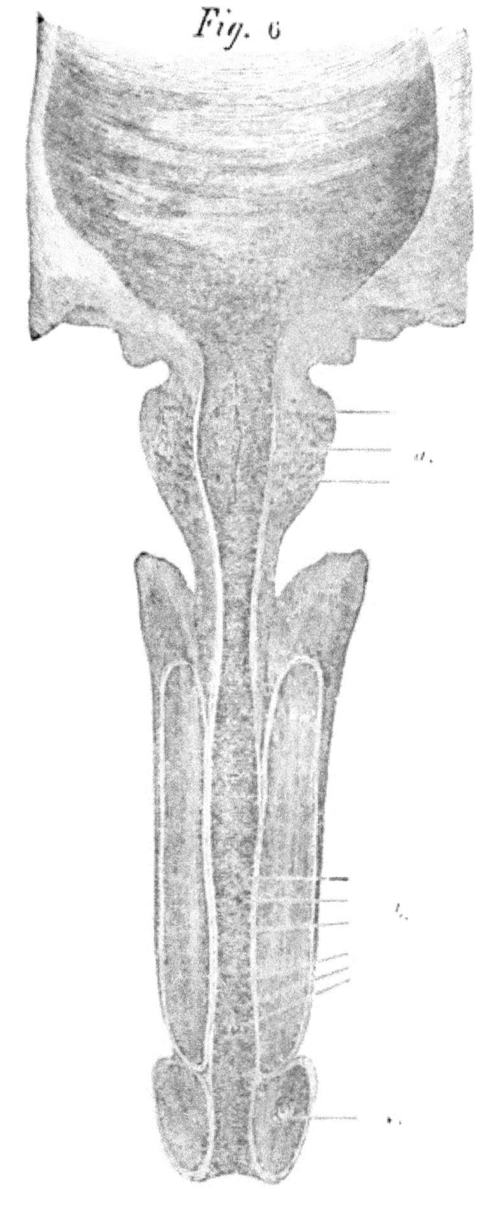

because the glans belong entirely to the cavernous body of the urethra; and if that body be paralytic or weakened from any preceding or existing cause, which we have often known to proceed from unnatural practices; in all those people where the spongy body of the urethra is not distended, impotence will arise, which, if not perfectly understood, cannot be cured by any physician; whereas, in healthy men, when these organs are in due tone during the orgasmus veneris, or the moment before the semen is ejected, the glans and whole cavernous body of the urethra are extremely turgid, so as to be ready to burst; but soon after, a kind of convulsive motion follows, and the semen is discharged with a slight loss of strength for a little time throughout the whole body, which soon recovers its usual vigor.

During coition, the corpus spongiosum and glans penis are rendered turgid by the blood filling their vascular structure, and the whole of the urethra is lengthened, but made narrower and straighter. The seed is gradually deposited in the sinus of the bulb, the glans being placed at the other extremity of the corpus spongiosum, and endowed with a peculiar sensibility. When a sufficient quantity of semen is collected, it excites the muscles covering the bulb to action, and the contraction of the fibres taking place, the semen is propelled rapidly along the canal; the blood in the bulb is at the same time pressed forwards, but requiring a greater impulse, it forms an undulatory wave behind the semen, narrowing the urethra, and urging on the semen with increased force.

When the passion of desire does not exist, the blood is not poured out into the cells of the corpora cavernosa, but returns by the veins as usual, and the penis remains flaccid; but when a person is under the influence of particular impressions which excite the nerves of these parts, the minute arterial branches, which before had their orifices closed, have their action suddenly increased, and pour from their open mouths the blood into these cells, so as to distend them, of course overcoming the elastic power that, under ordinary circumstances, keeps them collapsed. In this way the penis is rendered fit to convey the semen to the female organs of generation. The erection of the penis is greatly aided by the action of certain muscles, called the erectors of the penis.

The great vein of the penis is formed by branches from the gland, sides of the corpus spongiosum, and common integuments, runs along the back of the penis in the upper groove to its root, where it divides into two vessels which pass under the arch of the pubis, receive other veins from the prostate and bladder, and empty themselves into the internal iliac. The absorbents of the penis are very numerous, and terminate in the glands of the groins. A few observations here on puberty, and the changes it effects in the system, will, I think, be most opportune.

The approach of puberty, induces marked changes in the general system of the male and female, as well as in the local organs which are subservient to generation. The growth of hair on the chin, upper lip, and sides of the face of the male, and on the pubes of male and female; the peculiar alteration of the voice, especially in the man, the greater firmness of muscle, the extraordinary change in the passions and feelings, together with the great increase in the size of the penis and testicles of the man, show the advent of a peculiar change in the system, by which it is adapted for the propagation of the species. The desire for coition, implanted by nature for a wise purpose, becomes developed after the period of puberty, and the organs by which the act is performed, gradually assume their full vigor and dimensions.

The age at which the peculiar changes in the organism called puberty takes place, varies in different climates and in different constitutions. It is also influenced by the mode of life and circumstances of the individual. The period of puberty occurs earlier in warm than in cold climates; in temperate countries, it takes place from the fourteenth to the seventeenth year; the passions of youth living in large cities and towns, are, however, excited earlier than are those of the agricultural population, on account of the greater sources of temptation to which they are exposed.

In those animals which are not endowed with reason to guide their actions, the desire for copulation occurs periodically, and in some the testicles increase in size until the season of procreation is over, and then decrease, and continue small, until the commencement of the next season. Evidence of this may readily be found in the testicles of the cock-sparrow, which progressively increase in size from January till the end of April, when the love season of these birds usually terminates. The increase and diminution of these organs, however, do not take place in birds only, but has been discovered in many other animals, more especially in the land-mouse and mole.

There are several reasons which might be alleged

for the existence of a periodical desire for copulation among animals—were it otherwise, as the passion for sexual intercourse is very powerful, and animals do not possess the light of reason so as to be enabled to restrain or subdue their passions, it is probable that from its excessive indulgence, all their other habits might be lost, and even the necessity of providing for their present and future wants might be forgotten; besides which, in those animals which are very fruitful, and which do not long carry their young, their number would be in a short time exceedingly great, far beyond the means of support that nature has provided for them. Another reason might be alleged, that, were domestic animals always in heat, they would be of comparatively little service to man, while the flesh of wild ones would be too coarse and rank, and altogether unfit for the purposes of nourishment. The period of the year during which the desire for copulation principally exists in animals, is that of spring—few experience any sexual desire during the winter, except the frog, wolf and fox; the severity of the cold seems to destroy, at least for the time, all such feelings. On the other hand, in climates where the summer is very hot, the genital organs of animals become so much relaxed in tone, as to render them unfit for the proper performance of the necessary act. The case is, however, somewhat different in domestic animals; the passion is less periodical, the secretion of semen, not being arrested by cold, to which they are much less exposed, and the circumstances in which they are placed being altogether different.

Generally the desire in the male and female for procreation arises at puberty, and may be indulged in, if health and the requisite powers continue, at all times and seasons of the year. Being endowed by nature with the high, and exalted function of reason, they are left free agents, having the full power to use or abuse such capabilities, with the consciousness that if they do abuse the functions with which they are gifted, they must abide the penalty. Man is not affected by changes of temperature as are the wild animals, either as respects excessive heat or intense cold, and, consequently the human testicles are generally the same in dimensions after puberty throughout the year. The secretion of semen by man, begins about the period of puberty. The passion for copulation, and the secretion of semen, are indications of the great change which takes place in the system at that time. Those eunuchs only are not influenced by the desire for procreation who were deprived of the organs of generation prior to puberty; those who were castrated subsequent to that event, still entertain the desire for intercourse, although in a less degree than men who have all their organs entire. Desire is more languid in advanced age, than during the period of the adult life; the semen is then more sparingly secreted, and, indeed, all the functions of the system are performed in a less energetic manner, although, as will soon be shown, old men are not in every instance deprived of the power of generation. Desire is also generally moderate in persons who have small organs, occasionally it is altogether absent.

To the use of the sexual organs for the continuance of his race, man is prompted by a powerful instinctive desire, which he shares with the lower animals. This instinct is excited by sensations; and these may either originate in the sexual organs themselves, or may be excited through the organs of special sensation. Thus in man it is most powerfully aroused by impressions conveyed through the sight or touch; in many animals the auditory and olfactory organs communicate impressions which have an equal power; and it is not improbable that in certain morbidly excited states of feeling, the same may be the case in ourselves. Local impressions—as stated in the subsequent part of this work—has a very powerful effect in exciting sexual desire, as the experience of almost every one will attest; local diseases, as hereafter mentioned, often cause the most criminal acts. The instinct, for sexual intercourse, when once aroused, even though very obscurely felt, acts upon the mental feelings, and thus becomes the source, though almost unconsciously so to the individual, of the tendency to form that kind of attachment towards one of the opposite sex, which is known as *love.* This tendency cannot be regarded as a simple passion or emotion since it is the result of the combined operations of the reason, the imagination, and the moral feelings; and it is the engraftment, so to speak, of the physical attachment upon mere corporeal instinct, that a difference exists between the sexual relations of man, and those of the lower animals.

FEMALE GENERATIVE ORGANS.

The female organs of generation are classed in two divisions—External and Internal. The External consists of the *mons veneris, labia externa, perineum, clitoris* with its *prepuce, nymphae, vestibule, meatus urinarious*, *hymen* in virgins, and *carunculae myrtiformies* in matrons.

The Internal are the *vagina*, *uterus*, *two ovaries*, and *two fallopian tubes*. The latter four are strictly copulative, and the others, generative organs.

THE VULVA.

The *vulva*, or pudendum, is a collective designation for the external female genitals, including the mons veneris, the internal and external labia, the clitoris, and the orifice of the vagina.

The *mons veneris*, which is the same as in man, is the elevation of the integuments directly over the pubis. It is constituted of condensed cellular tissue and adipose matter, and in the adult is covered with hair.

The *labia externa*, or *majora*, are two folds of skin continuous with the mons veneris; they extend in the longitudinal direction, and terminate below in an angular commissure,—the *fourchette*. Like the mons veneris, they are composed of cellular tissue and fat, and form the anterior boundary of the sexual organs. The internal lining of these parts is a delicate, vascular epithelium, which is furnished with mucus follicles. The distance between the fourchette and the anus constitute the perineum and is about an inch in length.

The *labia interna* or *minora*, called also the *nymphæ*, are duplicatures of the epithelial membrane within the external labia. They arise from the anterior commissure of those bodies, and surround the base of the clitoris, thus forming the *preputium clitoridis*. At the inferior margin of the clitoridis the folds of the nymphæ unite in the *frænu-*

lum, from which point they diverge, and are gradually lost on the inner surface of the *labia majora*, at the orifice of the vagina. The *clitoris*, in several respects, resembles the male penis. It is formed of erectile tissue, and arises by two crura from the rami of the pubis and ischium. It has also a suspensory ligament, but is imperforate, and seldom exceeds the length of half an inch. Its free portion is called the *glans* clitoridis; its internal structure is similar to that of the corpora cavernosa penis, and, like the latter, also, it is an erectile tissue, acted upon by a small muscle, the *erector clitoridis.* These muscular fibres arise from the rami of the pubis and ischium, pass on the under surface of the clitoris, and terminate at its apex.

The *orifice of the urethra* is within the vulva, about an inch behind the clitoris, and directly above the vagina, where its position is marked by a small tubercle.

The *orifice of the vagina* is an elleptical opening below or behind the meatus urinarious, having a thickened margin, and being founded laterly by the nymphæ. It is usually more or less closed by a duplication of mucous membrane, the hymen, which exists sometimes as a perforated septum, and sometimes as a semilunar fold. It is more rarely imperforate; in other circumstances, it has two or more orifices, and again it is represented by a partial, fringed margin, or may be congenitally deficient. Its rupture leaves an irregular edge, with thickened cicatrices, known as the carunculae myrtiformes.

The vestibulum is the triangular surface between the clitoris before, and the vagina behind; the urethra opens into it, and it is bounded latterly by the labia interna.

PLATE 6.

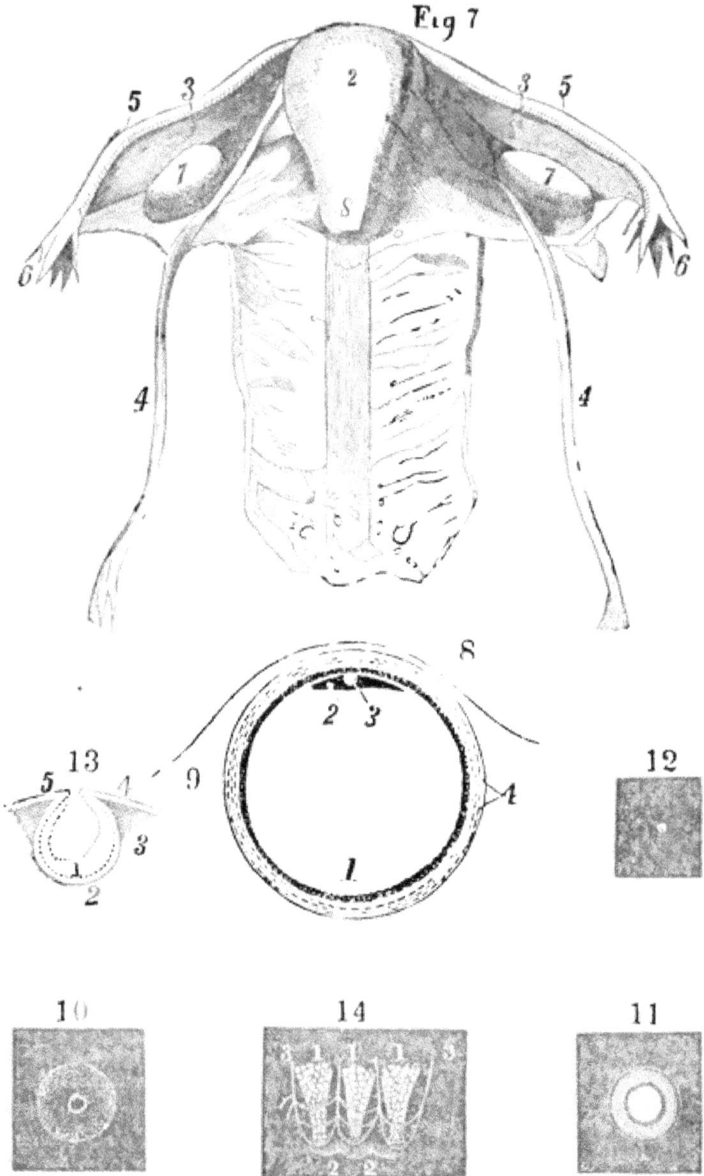

Explanation of Plates.

Fig. 7, Represents the vagina laid open by a longitudinal section, with the uterus and its appendages *in situo.*—1, Orifice of the vagina with the carunculæ myrtiformes; 2, Body of the uterus; 3, 3, Broad ligaments; 4, 4, Round ligaments; 5, 5, Fallopian tubes; 6, 6, Their fimbriated termination; 7, 7, Ovaries; 8, Os tincæ. Fig. 8, Diagram of a section of the unimpregnated graafian vesicle and its contents, showing the situation of the ovum.—1, Membrana granulosa; 2, Proligerous disk; 3, Ovum; 4, Inner and outer layers of the wall of the graafian vesicle; 9. Indusium of the ovary, derived from the peritoneum.

Fig. 10, The unimpregnated ovum surrounded by its proligerous disk, magnified about fifteen diameters. Fig. 11, A section of an unimpregnated ovum, representing the thick external envelopes, connected with the inner surface of the latter, the germinal vesicle. Fig. 12, Represents, within a sqare area, the unimpregnated human germinal vesicle magnified forty-five diameters. On one side of it is seen the germinal spot. Fig. 13, Is a corpus luteum, taken from a female who destroyed herself by drowning, eight days after impregnation.—1, Mucous tunic of the graafian vesicle sprouting from the circumference toward the centre; 2, External tunic of the vesicle; 3, Ovarian stroma; 4, Ovarian membrane; 5, Point at which the ovulum escaped from the graafian follicle.

Fig 14. Represents the interior of the uterus, with the germ in progress of formation.—1, 1, 1, shows them much enlarged; the matter which is to become the *membrana decidua,* is seen deposited between and upon their surface; 2, 2, 3, 3, Uterine arteries extending into the decidua and there forming loops.

General Observations on Remarkable Cases.

Seminal debility—involuntary loss of semen—will cause the following complications, in seven-eights of the afflicted, when it, or the practice of self-abuse, has existed for two or three years.

These extracts are from letters of patients, describing their symptoms, when they applied to us for treatment, and all of whom we are happy to say, we have since cured, though some of them were under our care a year, before we would consent to its discontinuance. We relieved nearly all of such patients, however, of the painful and troublesome symptoms in a very few days. Patients with diseases of this nature, will experience a relapse if the treatment is not persevered in for a considerable time after they have been cured. We therefore take this occasion to warn them from doing so, till informed by us their cure is permanent, as the disease is much more obstinate after such relapse.

As all of our communications are strictly private and confidential, we, of course, neither attach, nor copy the addresses, but if any persons—who may wish us to cure them—doubt their genuineness, we will present them with indubitable proofs, upon a personal visit. All patients may place implicit confidence in the strictest secrecy being kept by us, as we never divulge the name or residence of any patient, upon any pretext whatever.

THE VAGINA.

The vagina is a canal interposed between the uterus and the vulva, and between the bladder and the rectum. Its length is from four to six inches; it is somewhat dilated near the uterus, and contracted at its commencement in the vulva, between which points it curves obliquely forward and downward, being flattened transversely, and having its parietes in close apposition. It is composed of three laminæ; the external one is a fibrous structure, strong and contractile, and much resembling the dartos of the scrotum. The *middle layer* is constituted of erectile tissue enclosed between two layers of fibrous membrane: it commences directly within the vagina, surrounds its upper half, and extends about an inch backwards in the direction of the uterus, its thickness being seldom more than two lines. This is the *corpus spongiosum vaginæ*, so called on account of its resemblance to the analogous structure in the urethra. It is also called the *plexus retiformis*, and embraces a remarkable congeries of veins.

The *internal membrane* belongs to the mucus class, it is furnished with laminæ, and is marked by transverse folds or ridges—the rugæ of the vagina—which, however, are chiefly confined to its anterior portion. A raphæ, or longitudinal slightly elevated line commences on the anterior parietes of the vagina just within its orifice, and extending back is lost near the uterus, and a nearly similar line exists on the posterior surface. The epithelial covering of the mucus membrane is remarkably developed, and presents an abundant papillary structure.

The *glands of Duverney* resemble, both in form and function, the glands of Cowper in the male.

They are situated on either side of the orifice of the vagina, beneath the sphincter muscle and the superficial porineal fascia. They are from half an inch to an inch long, narrow and flattened, and their excretory duct opens in front of the carunculæ myrtiformes.

The sphincter vagina muscle has been already described. It embraces the orifice of the vagina like a broad ring, and covers the plexus retiformis.

THE UTERUS.

The *uterus* or *matrix* is a pyriform body, suspended in the pelvic region between the bladder before and rectum behind. It varies greatly in size; its average length is about two and a half inches, its breadth across the upper and widest part an inch and a half, and its thickness an inch. It is divided into the fundus, or that part above the Fallopian tubes; the cervix or narrowed portion below; the body, which is intermediate between the fundus and cervix, and the ostincæ or opening into the vagina. It is convex behind and flattened in front, with its base directed upward and forward, and its neck downward and backward.

The anterior surface of the uterus has a peritonial covering over its upper half, which membrane is reflected from it to the bladder, constituting the anterior ligament. The posterior surface is entirely invested by the peritoneum, which in this instance, is reflected upon the rectum, so as to form a pouch between, and is called the posterior ligament. On the sides of the uterus, the anterior and posterior laminæ of peritoneum, meet in a longitudinal median line, and on them reflected from it to form the *broad ligaments.* These ligaments are extended transversely to the parietes of the lesser pelvis on each side, and thus form a paritial septum in its cavity, and transmit between their laminæ, the blood-vessels and nerves, the uterus and ovaria.

The *round ligaments* are given off on each side from the body of the uterus directly beneath the Fallopian tubes, whence they pass upwards and outwards until they reach the internal abdominal ring; they then run through the abdominal canal, come out at the external ring, and are inserted into and lost upon the mons veneris. They are strong, fibrous vascular cords, embraced between the folds of the broad-ligaments; and they are accompanied in the abdominal canal by a process of peritoneum, which sometimes extends to the exit of the ligament at the external ring.

The *cervix uteri* projects by a mammillated extremity into the vagina, and the two are joined at the base of this prominence. The cervix has a central perforation,—the *os tincæ* of an oval shape, and placed transversely. It is bounded in front by a prominence of the cervix, called the anterior lip, and behind by a similarly formed but much smaller projection—the posterior lip.

The *cavity of the uterus* is triangular with an orifice at each angle; one of those three orifices and much the largest, is continuous with the canal of the cervix, and of course communicates directly with the vagina, and is called the *internal orifice* of the uterus. It is smaller than the *os tincæ;* but the canal between those openings is somewhat dilated, flattened from front to back, and marked by two longitudinal lines, one on the anterior, the other on the posterior surface. These are again crossed by transverse and erratic striae, named from their branched appearance the *arbor vitae*, among which are interspersed some small mucous glands and follicles, known as the *ovula Nabothi* They are not confined to the cervical canal, but are also found in the cavity of the body. At the angles in the upper part of that cavity are two small orifices, one on each side, opening into the Fallopian tubes.

The *uterus* is composed of three very dissimilar tissues, an exterior serous covering derived from the peritoneum; an internal mucus membrane, and between these the proper structure of the organ.

The *peritoneal coat* has been already noticed, together with the manner in which it forms several of the uterine ligaments. The *mucus* membrane is a continuation of that which lines the vagina and is continued into the Fallopian tubes. It can only be separated in patches and with much difficulty; whence its existence as a separate membrane has been sometimes denied.

The *proper tissue* of the uterus is of a greyish color, of an almost cartilaginous hardness, and composed of fibres, respecting which a great diversity of opinion has always existed. By some anatomists they are regarded as simple fibrous tissue; by others, as muscular structure; and by others as a convertible substance, that is fibrous in the unimpregnated state, but which, during pregnancy, acquires the character of muscular fibre, resembling that of the viscera of organic life, and like it endowed with contractility. Such is the opinion of Cruveilhier, who remarks, that the great influx of blood into the uterus and the consequent distension and development of its fibres in the gravid state, reveal a structure which was before concealed by an atrophy consequent to inaction; and he adds, that this view is confirmed by the microscopic observations of Roederer, the chemical experiments of Schwelgue, and also by the results furnished by comparative anatomy, which has shown circular and longitudinal muscular fibres in the uteri of some animals even in the unimpregnated state.

The gravid uterus, however, presents two layers of fibres which have the structure and function of true muscle: the *external layer* is formed both of longitudinal and oblique fibres, the former being chiefly developed on the anterior and posterior surfaces, and upon the fundus; the oblique fibres are most distinct at the sides, where they are continued upon the Fallopian tubes and the round ligaments. The *internal layer* consists of circular fibres, with some longitudinal fibres on their internal surface; but around the cervix this structure becomes annular, the fibres decussating at various angles. The muscular character of the uterine body, in pregnancy, is strongly evidenced by its powerful contractions in the process of parturition.

The internal substance of the uterus is thus composed of a mass of closely interwoven muscular or fibro-muscular fibres, together with a vast congeries of blood vessels, which though small in the unimpregnated state, acquire a surprisingly augmented size during the later periods of pregnancy.

The *arteries* of the uterus, four in number, are derived from two sources: two from the hypogastric trunk, forming the proper *uterine arteries*, and two from the ovarian arteries, of which branches are also distributed to the ovaries and Fallopian tubes. The uterine veins are of great size during pregnancy, and are then called the *uterine* sinuses. The nerves of the uterus are derived in part from the hypogastric, in part from the renal plexus.

THE FALLOPIAN TUBES.

These trumpet-shaped canals are given off from the superior angles of the uterus. They are embraced within the folds of the broad ligaments, being undulated in their course and variable in their diameter. They are four or five inches in length, and extend almost to the sides of the pelvis. Their diameter at the uterine orifice will only admit a bristle, but the canal at its external or free termination is as large as a quill. This outer end is broken into a triple series of fringe-like irregular processes of unequal length, constituting the *fimbriated* portion, or *corpus finebriatum*, in the centre of which is seen the orifice of the tube called the *otium* abdominale. One of the processes is attached to the proximate part of the corresponding ovary, by which means these structures are retained in their relative position.

The tube is in itself a strong fibrous cord resembling the tissue of the unimpregnated uterus, and it is invested by the peritoneum by being placed between the duplication of this membrane that forms the broad ligament. The internal coat is mucus, analogous to that which lines the uterus, and remarkable for presenting a gradual transition into the peritonial coat, with which, at its fimbriated orifice, it is continuous; in other words, the mucus terminates in a serous membrane, and thus in the two openings of the Fallopian tubes exist the only two normal perforations of the otherwise perfect sac of the peritoneum.

The use of the Fallopian tube is that of an *oviduct.* It receives the ovum from the ovary and transmits it to the uterus, and also provides it with a double envelope—an internal one of a gelatinous or albuminous nature, and an external one, the *chorion,* which is fibrous, and appears to be produced by the exudation of fibrine from the lining membrane of the tube.

THE OVARIES.

These bodies, one on each side, are placed within a duplication of the broad ligament, and behind the Fallopian tubes. They are of a flattened oval form, reddish-white color, and unequal fissured surface. They are retained in their position partly by the broad ligament, and partly by an *ovarian ligament,* a rounded cord that connects them with the upper angles of the uterus, below the Fallopian tube. The ovaries are about an inch in length, and their distance from the uterus is about an inch and a half. They have an external investment derived from the peritoneum; within which, and closely adherent to it, is the proper fibrous capsule, strongly resembling the tunica albuginea of the male, and, like the latter, sending prolongations into the gland that divides it into irregular compartments resembling a network of areolar tissue. Within and lining this fibrous capsule is a vascular membrane, analogous to the tunica vasculosa in man. The fibrous and vascular tissues are intimately blended into a spongy mass, called *stroma,* in the midst of which are the *Graafian vesicles,* of which bodies ten or fifteen exist in a mature state in each ovary, besides a vast number that are imperfectly developed and never reach the perfect state.

The *Graafian vesicle*, or *ovisac*, consists of two layers, of which the outer one is a mere vascular thickening of the surrounding ovarian stroma; while the internal one, which is the true ovisac, is transparent, and has no obvious structure. Within it is placed the *ovum*, the latent germ of the future being, between which and ovisac is a granular matter, arranged in the following manner: a series of the granules surround the ovum in a discoidal form, and assume the appearance of cells, so united as to form a sort of membrane, which is called by Dr. Barry the *tunica granulosa*, by others the *proligerous disk*. The granules that line the ovisac within are also collected in a membraniform structure, the *membrana granulosa*. These two parts are connected by four band-like extensions of the same *cellulo-membranous* structure, which seem to suspend the ovum in its place, and are called the *retinacula*. The space between the membranes which is not occupied by the retinacula, is filled with fluid, in which few or no cells can be seen.

The *corpus luteum* is a yellowish, spongy tissue, granular, friable and vascular, having a small central cavity lined by a delicate membrane. It is the cicatrix left after the escape of the ovum from the ovary, and consequently varies much in size according to the time which has elapsed since conception. At first it is large, bean-shaped, and prominent, so as to occupy from a fourth to a half of the ovarium. But after parturition it diminishes in size, and in a a few months a cicatrix alone remains, and even this is finally effaced. Whence it happens that neither the corpora lutea, nor their remaining cicatrices, are certain indications of the number of children a woman may have borne. Dr. C. D. Meigs has published an interesting memoir on the corpus luteum, in which he maintains that the apparent structure, form, color, odor, coagulability and refractive power of this body are similar to those of the yolk of the egg; a true *vitellary matter* "deposited outside of the inner concentric spherule or ovisac of the Graafian vesicle."

FUNCTION OF THE OVARIO-UTERINE SYSTEM.

The result of successful coition between the two sexes is the injection of a certain portion of the spermatic fluid into the vagina; and fecundation appears to consist in the direct communication of the male spermatozoa with the Graafian vesicle of the ovarium, through the fissure of the zona pellucida to the contained ovum,. which ruptures and escapes from the ovisac. The corresponding Fallopian tube simultaneously embraces, by its fimbriated extremity, the ovarian surface, and receives the detached ovum into its canal, while the ovisac itself remains as the lining membrane of the corpus luteum. While the fecundated ovum is yet in the Fallopian tube it acquires a gelatinous covering, the *amnion*, which is again surrounded by a membrane of fibrous texture, called the *chorion*. The amnion secretes a fluid, the *liquor amnii*, in which the germ is suspended. How long the ovum remains in the oviduct, in other words, what time it takes in its transit from the ovary to the uterus is not certain, but appears to vary from eight to fourteen days. The first action of the uterus is the secretion, on its inner surface, of a delicate cibriform membrane, the *decidua*, which is composed of two layers, the *decidua vera*, that lines the uterus, and the *decidua reflexa*, that covers the ovum. Next forms the *placenta*, which results from the penetration of the vili of the chorion into the structure of the decidua vera. Its *fœtal portion* is derived from the umbilical vessels, which diverge in every direction from the point at which they enter its substance; or, in other words, it is generated by the extensions of the vascular tufts of the chorion, formed from the capillary terminations of the umbilical arteries and veins. The *maternal portion* of the placenta is formed by the enlargement of the decidual uterine vessels, and these assume the character of sinuses, against which the fœtal tufts project so as to form out of it a sheath for themselves. The blood is conveyed into the maternal placenta by the uterine arteries, and is returned by the corresponding system of veins; but there is no direct vascular communication between the two placentae

PUBERTY.

The period of puberty, the commencement of that part of life which is distinguished by the capability of propagating the species, does not occur exactly simultaneously in the two sexes; and still greater variety in this respect is caused by difference of nation and climate. Puberty declares itself in the female sex of our climate about the twelfth, thirteenth or fourteenth year, sometimes later, and is indicated by the occurrence of menstruation. In the male sex puberty begins about the fourteenth, fifteenth, or sixteenth year, and is attended with the secretion of semen, and with the occurrence of discharges of that fluid. In hot climates the body undergoes the changes of puberty earlier than in cold climates. It is stated that in the hot regions of Africa, they take place in the female sex as early as the eighth year, and during the ninth, in Persia. Young Jewesses are said to menstruate earlier than other females. The capability of reproduction generally ceases in the female sex, together with the function of menstruation, between the forty-fifth and fiftieth years. The duration of the reproductive power in man cannot be so exactly defined; in general, it continues longer than in woman, and not unfrequently very old men manifest a remarkable degree of virile power.

The changes in the system which characterize the period of puberty, are partly local, affecting the generative organs, and partly of a general nature. The local changes consist in the growth of hair on the pubes in both sexes; in the menstruation of the female; in the copious formation of semen, and occurrence of erection in the male; and in the enlargement of the breasts in the female sex. The general changes of the system affect principally the respiatory and vocal apparatus, the entire form of the body and the physiognomy, the character of the mind, and the feelings relating to the sexes. The respiatory organs acquire an increase of volume at the age of puberty, especially in the male sex; and the vocal apparatus undergoes the change so readily noticed in boys of sixteen to eighteen, the shrill, squeaking, broken and ultimately the deep-toned masculine voice. The whole body attains its most perfect form; while the features receive their stamp of individuality, and present signs serving to express the passions, though they are not as strongly marked as in many adults. Boyish indifference is changed to the most marked attention to the opposite sex. Sexual ideas arise instinctively and obscurely in the mind, and set in action the creative power of the imagination, but, at the same time by their influence on the whole mind, call into play the noblest mental faculties, so as to elevate and adorn the feeling of love.

MENSTRUATION.

Menstruation is the periodical discharge from the female generative organs of a bloody fluid poured out by the inner surface of the uterus. The first discharge is usually preceded and accompanied by some symptoms of general disturbance of the system, namely, by abdominal congestion, pain in the loins and a sense of fatigue in the lower limbs. Its periodical return is also attended in most women by unusual symptoms, which vary in different persons. The menstrual periods occur usually at intervals of a solar month; their duration being from three to six days. In some women the intervals are as short as three weeks, or even less; while in others they are longer than a month. Aristotle made the extraordinary statement, that menstruation rarely occurs every month; but in most women only every three months. This is evidence that the sage philosopher was an old bachelor, and studied books and not nature. I might as well add here, that those absurd and mischievous books purporting to have been written by Aristotle, are full of as false statements, silly directions, and are blind leaders of the blind.

Menstruation does not usually occur in pregnant women, nor in most cases in those who are suckling. In rare instances, however, it continues during the period of suckling, and even during pregnancy.

The sun, moon, stars, nor even the migratory comet, do not exert an influence over this function, for upon every day in the month different females have their catamenia. It is now well known that the menstrual periods tend to regulate and moderate sexual indulgence. This will be more fully explained under the head of *Conception.*

SEXUAL INTERCOURSE.

All the phenomina connected with the sexes which animals present, are dependant on the formative generative organs; the ovaries and testes, and on the influence which they exert on the rest of the organism. Not merely does castration during youth, for the most part, prevent the development of the sexual feelings and emotions, but even when performed at adult age, that operation destroys almost entirely the sexual excitability. This is also the case with animals, in which common observation detects its effects.

The almost entire physical change which occurs in animals after this operation is readily suggestive of its controlling influence over the body as well as over the mind of men and animals.

In both sexes the act of coition is attended with pleasurable sensations, but their respective share in the act itself is very different. In the female there is but a partial expenditure of the nervous power, when she fully participates in the sexual act, and none scarcely when her feelings do not prompt her to return the ardor of the male; neither does she expend any nervous power in the production of erection; no energetic rythmic muscular contractions when the venereal excitement has reached its height, and no emission of semen; but merely an increased secretion from the vagina, excited by the impressions on the sensitive nerves of the female sexual organs, and serving to lubricate the passage to facilitate sexual commerce.

The man feels exhausted after the act; the woman simply yields herself up to the pleasureable excitement. The clitoris, which is known to be the part most susceptible of the pleasurable sensations in in females, is not like the penis of the male, rendered by friction the seat of intense sensation and nervous excitement during coition, and hence its excitability is found not to be wholly exhausted after the act is completed.

SEMEN.

The semen, or fecundating matter, contains the property which gives integrity to the germ, and is capable of determining the particular form of the new animal or plant, but it is defective and cannot develop that form until it has united with the female germ. The defects of the ovum, or germ, and of the semen, are not the same in nature, for each contains that part in which the other is deficient. The ovum and semen are not similar halves of one whole. The ovum of animals contains the part destined to germinate, and is, in fact, the primary particle which forms the basis of the new organism, and maintains uniterrupted the chain of organization. The semen, on the contrary, does not itself germinate, but is a fluid excitor of germination, endowed with the power of determining the form, not only of the species, but of the individual organism which produced it.

CONCEPTION.

Pregnancy may occur at any period after menstruation has been established, and until the cessation of that function.

That conception occurs more readily immediately after the menstrual periods, is a remarkable fact; but that it will not take place after eight or twelve days from that period, as stated in some books, written to misguide, or from ignorance, is absurd and untrue.

The catamenial function, is nature's guarantee that the females in whom it exists, are capable of conceiving and bearing offspring.

The conditions necessary to the consummation of this object is, that the semen or seed from the opposite sex should enter the womb. Quantity has nothing to do with the success.

The grand office of the semen is to awaken the life which previously exists in the animulculæ in the ovaries of the female, and constitute (so to speak) the characteristic principles of vitality of both into one individual embryo.

To effect successful commerce between the sexes nature has arranged the organs of generation to a perfect adaptation, and impressed all animals with the desire to cohabit. The reward of obedience to these natural laws are perfect physical happiness and the procreation of the species.

In man and mammalia fecundation is the result

of the successful union of the sexes, in which the semen of the male enters the womb, traverses the Fallopian tubes to the ovum, and then comes into direct contact, in order that the function of impregnation be matured.

This has been proved beyond the possibility of a doubt, by direct experiments upon numerous animals with uniformly the same result.

It has been frequently observed that men whose copulative organs are very short, or those in whom the malformation consists of the opening or mouth of the penis terminating any considerable distance from the end of the organ, are rendered impotent; from the fact simply that the condition of the penis prevents the semen from being thrown directly into the womb.

PASSAGE OF THE OVUM INTO THE WOMB.

The human ovum reaches the uterus in from one to two weeks after impregnation. The embryo is enclosed in a membraneous sac which nature has formed for nourishing and protecting it. When the blood vessels of the embryo have reached its surrounding membranes which have been developed from cells, and participate in the active properties of cells, it absorbs the nutriment. The nutritive matters thus absorbed are supplied by the blood of the mother in the human placenta. The absorbed nutriment enters directly into the blood of the fœtus. The process thus maintained between the fœtal and maternal blood, supplies the process of respiration tq the fœtus, or an equivalent for that process.

PLATE 7.

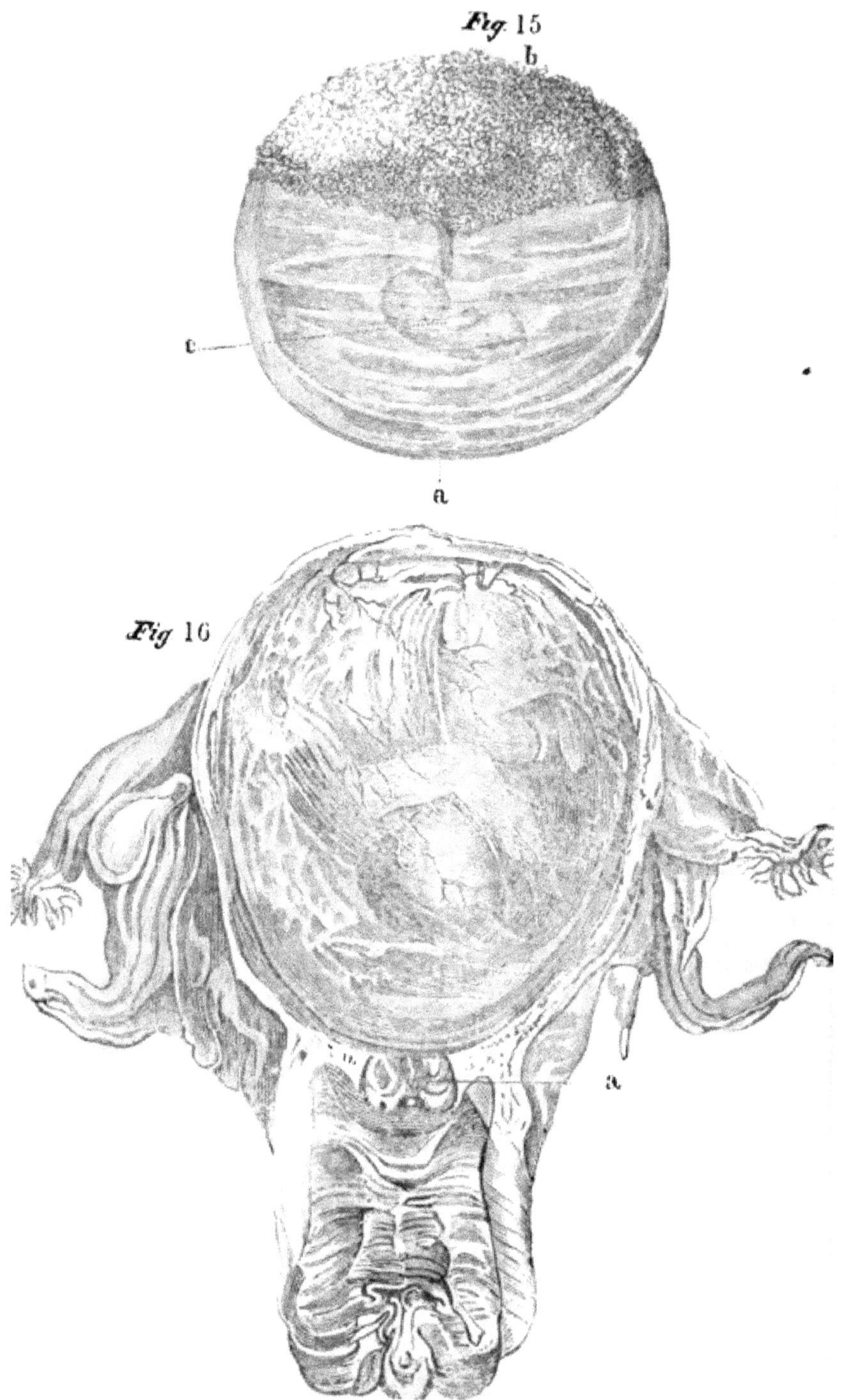

Explanation of Plate 7 and Figures 15 and 16.

FIG. 15.—Represents a fœtus of two months' gestation (growth), surrounded by its uterine membranes, attached by the umbilical cord to the placenta (after-birth), and floating in fluid (amnial), and is the most perfect one of this period I have ever seen; (*a*) the pellucid membranes, (*b*) the accumulation of vessels forming the placenta; (*c*) the embryo seen through the membranes.

FIG. 16.—Displays a fœtus of five months within the womb, which has been laid open; blood vessels are seen running through the various textures, and the cord is seen twisted around the neck and left ancle. Below is seen the postern surface of the vagina, and on each side of the womb is seen the fallopian tubes terminating in many finger-like projections, and between them and the womb the ovaries.

REMARABLE CURES.

June, 1858.

SIR:—As we have arranged the fee for my cure, now for the desired explanation. Without having heard of Masturbation, or its consequences, and wholly unsuspicious of them, I did indulge to some extent in it. I should judge between 14 and 17 years old. I am now 55. It is quite likely to excess for some part of the time, for at first I thought no evil in connection with it, but as I grew older, I knew better and abandoned the habit. The result of this involuntary loss of semen continuing, was to produce a state of atony for some years—an almost total loss of virile power,—though from the age of 30 I have been able to partially cohabit at times, when under the treatment of some of the best physicians. The urethra has been sore ever since that period. This sub-inflammation (I suppose) now exists. It is for you now to infer the nature and extent of the functional derangements I labor under. Their indications at present are a loss of *virile power*, a weak flow of *urine* generally breaking to pieces as it falls,* occasional pains within the *rectum*, and a sore feeling in voiding urine. I have thought at times, that sexual commerce abated the seeming soreness of the parts. The feces when costive, are covered with blood and matter on one side, pain in micturation sometimes, and a few drops of creamy looking matter will follow the urine. I am very often obliged to get up in the night to pass urine; sometimes very restless, and unable to sleep; linen discolored, and sometimes I have a yellow discharge from the urethra, and in the morning shredy substances pass with the urine. I have a pain at night in the *rectum*, which will last for hours, and be so severe as to prevent the passage of the urine."

This case presents nearly the same complication and symptoms, as almost every patient suffers with, whose disease has existed for any great length of time. Many young patients even at this writing though they have nearly all such symptoms, have not attained the 20th year. Out of the number of about one hundred and fifty patients of this class, under our care and treatment at one and the same time, more than one hundred of them applied to us with these seminal diseases, complicated with severe and dangerous affections of the urethra, testicles, bladder, and kidneys. Nearly all of them are laboring under indigestion and costiveness, and their nervous system so affected as to be partially or totally unfit for study or business. The affection of the bladder and kidneys is a most distressing and dangerous complication, as any physician will inform you.

DEVELOPEMENT OF THE FŒTUS

Within seven days after successful copulation and conception, the ovum (female germ) envelopes itself in membranes and passes into the womb through the Fallopian tube. The womb has also generated membranes within its cavity for the reception and nourishment of the embryo.

The membranes enclosing the embryo, secrete a watery fluid in which it floats, protected by this means from uterine pressure and injury, or destruction from concussion or jars of the body of the mother.

So small and delicate are the embryonic structures that at the commencement of the second month of gestation, the length is only from a few lines to half an inch.

The extremities are leaf-like appendages and just visible. Then the mouth also exists, soon after the anus is seen ; the last bone of the spine, (the coccyx) is very large and the first bone formed. The head acquires a considerable size ; the eyes advance forward from being lateral, to their natural anterion position, and the nasal cavities are soon apparent. During the second month, the umbilical cord is enveloped in its sheath, and the intestines are discernable. About the end of the second month, the bones and muscles begin to form ; the heart is covered, and the great arteries assume their permanant position. The glandular organs, the lungs and liver exist ; then appear rudimentary kidneys and the testes (Testicles) or ovaries. The external organs of genera tion make their appearance in the shape of wart-like prominence ; the bladder soon after makes its appearance.

Up to this period the mouth and nasal cavities are not separated ; the eye-lids and external ear are now apparent, and the extremities are more distinct, the fingers and toes are marked by separation. At this period of developement the embryo is about an inch long.

In the course of the third month the fœtus acquires about three inches in length and proportionate completion of developement. In the fourth month the sex is perceptible, and it has obtained four inches in length, and in the fifth twelve inches; at this period begins the formation of the fat and nails, and signs of fœtal life are perceptible to the mother. The fœtus born at the sixth month of gestation often breathes but does not live. During the seventh month the fœtus has acquired about sixteen inches, and if born may live. The skin is red. In the eighth lunar month the eyelids become free, and the testes descend from the belly to the scrotum. In the ninth month the hair appears on the head and the embryo measures about seventeen inches; at the lower part of which it increases in size, so that in the tenth lunar month it measures about twenty inches, and weighs from six to twelve pounds

The weight of infants at birth vary very much, but the first child a mother weans is generally smaller than those born subsequently·; and males weigh more than females, being a trifle larger also. I have myself seen a child live a few days, which weighed three and a half pounds. The elder Dr. Ramsbotham confined a woman who was delivered of a child weighing sixteen and a half pounds, and Mr. O. D. Owens assisted in the delivery of a child weighing seventeen pounds twelve ounces, measuring twenty-four inches in length. This was the largest child ever delivered, which is on record.

Birth occurs at the end of nine months.

STERILITY, CAUSES AND CURE.

The principal end of man's earthly existence is offspring. However widely opposed their general natures may be, all are united in the desire which finds its accomplishment in the reproduction of their kind. I say all, for I do not hesitate to assert without fear of contradiction, that no man and wife ever had their affections perfected or developed who were deprived of legitimate offspring. It is one of the first duties of the human race to increase and multiply, and the man who leaves the world without having obeyed the injunction, can scarcely be said to have fulfilled the great end of his existence. All men who are not of monstrous conformation, or who have not been seriously injured by artificial means, are equal to the task of reproduction; indeed, without the parts and means necessary to reproduction, he could scarcely exist at all, and would be no more a human being than if he were deficient of heart or brains. There is no such thing as barrenness in *natural* women, and the causes which are supposed to render women so, can, in ninety cases out of a hundred be removed.

It is not denied, however, that a great many married persons are unblessed with offspring, whose exertions are undoubted, and who would give much if it were otherwise. The causes of unfruitful marriages are numerous. One is, their unfitness in consequence of malformation, besides Fluor Albus, (generally termed whites,) Leucorrhœa, Prolapsus Uteri, (falling of the womb,) Chronic Gonorrhœa, or Gleet, extreme indulgence, and a very vitiated state of the system of either, or both man and wife, from Scrofula. The obliteration of the vaginal canal, or absence of the ovaries or uterine tube in the female, also are causes; but these are rare occurrences. My opinion is, that but very few females are of necessity barren Young married persons almost invariably become physically adapted to each other, though it occasionally happens that a couple will have no offspring, and yet being divorced, will form other connections, and both will have children. It was so with Bonaparte and Josephine. A French writer says, that good authority reports the emperor as having used various tinctures, borax, marjorum, etc., of course to no effect. There is sometimes too much ardor in persons of full habits and amorous propensities; their intensity, however, is generally qualified by time. This is well exemplified in newly married couples, as the first year of their marriage is generally unfruitful from their too frequent amorous embraces, thereby preventing the semen from attaining a healthy state. Moderation may be produced by a light vegetable diet, cooling medicine, and occasional sea-shore trips. Conjugal enjoyment on the part of the female should be followed by repose, as but very little motion or agitation in persons of warm temperaments is sometimes sufficient to prevent the *ovulum* from reaching its proper location. A female desirous of conceiving, must not cohabit too often, for the first month after her supposed conception, as the spasmodic agitation consequent on the embrace of a very amorous couple, is calculated to disturb the embryo in its earlier state of existence, and hence occasion abortion or miscarriage.

Baillie, Swaumerdaur, Larry, Dubois, and others, say, that the great leading cause of sterility is weakness or debility on the part of the male or female, or both; and Dubois says, if this matter was duly attended to, nine-tenths of the people who are now pining for heirs might be blessed with numerous progenies.

Libertinism, and that horrid and loathsome practice of *self-abuse*, or *masturbation*, are the principal causes, and I have, therefore, treated upon them largely in another part of this work. Among the less numerous causes, are dancing immoderately, and tight-lacing, as the pressure causes a weakness and lassitude of the system. Stimulating drugs, etc., such as Cantharides, tincture of lyttæ, essence of marjorum, arrow-root, syrup of pine-apples and port-wine, mushrooms roasted and steeped in salad oil, or borax, are worse than useless, for many of them actually injure the person, as my large practice for many years too well prove. General physicians often administer such remedies, when they ought to know they will prove injurious; and if they do not, they are equally at fault; but those who should be eternally cursed are the unprincipled quacks, who palm off their injurious and nauseating stuff for money, regardless of the lasting injury to those who are unfortunate enough to use them. Yet some of these remedies sold for the cure of sterility, impotency, weakness or debilitation, diurnal and nocturnal emissions, or loss of semen—with the consequent impoverishment of the whole system—are actually recommended by physicians, either from ignorance or self-interest. I cure physicians every month of such diseases, which is the best proof possible to obtain, of their total ignorance of the causes, and proper treatment of such cases.

Some authors think there are fertilizing virtues in water-cresses, duckweed, carrots, dandelions, artichokes, figs, potatoes, shell-fish, peaches, hemp-seed, eggs, calves'-feet jelly, etc. The most of these are incentives to amorous propensities, but no farther. Ludwig says the females of some countries swallow spiders, flies, ants, crickets, and even frogs to promote fecundation.

Morning is undoubtedly the most auspicious to generation. When a female with a low womb is married to a masculine man, they must correct the difficulty by means that may seem obvious, or they will probably have no offspring, and the female suffer agony instead of pleasure. If the semen is placed beyond the proper location, it cannot impregnate. When the case is directly otherwise, a proper remedy will be necessary to secure it from falling short.

It is a popular error that there is a mode by which male or female offspring may be produced at will. It is of no consequence whose theory of the mysteries of reproduction is correct, they are agreed on certain points, which shows this to be impossible. There are tolerably conclusive rules, however, for telling the sexes of children before they are born. Ladies experience more sickness with boys than with girls, which may be caused by their generally being larger and more lively. Their appetites generally vary, such as food that is hearty for the one, and of a different kind for the other. A roundness of the form promises a boy; whereas, when the tendency is nearly all to the front, and the hips and back give but little evidence of the lady's situation, the great probability is that the little stranger is a girl When a pregnant female is prone to sickness in the morning—longs for food of an invigorating quality—and carries her increase of form rather all round her, than in any particular place, the chances are altogether in favor of a boy; whereas, if her symptoms are otherwise, and as described above, she will, in all probability, be delivered of a girl.

PREVENTION OF OFFSPRING.

While all must admit that the reproduction of our kind is the evident intention of the sympathy of the sexes for each other, it is equally certain that there are numerous cases in all countries, wherein such a consummation were better avoided. For example, indigent people cannot be very anxious for numerous offspring to rear up in poverty; very fruitful females must find it very unpleasant to be nearly always in a state of pregnancy, nor is it to be supposed that married persons who are afflicted with hereditary diseases, can derive happiness from bringing into the world beings whose existence may, in all probability, be a burden to them.

Many females are so constructed as only to be able to give life to others at the sacrifice, or, at least, the risk of their own. This consequence alone, if no other, should prompt the physiologist, physician and philanthropist, to seek some remedy which would avert such awful suffering. Another reason, people, under all circumstances, whether they are poor or afflicted with diseases, or so organized, as to risk life in reproduction, will get married, hence, anything that will prevent the evil complained of, and yet allow nature her full rights, cannot but be of incalculable benefit to the public.

Other medical writers have treated upon this subject, but from either ignorance or delicacy, have not done so with sufficient fullness. I will reiterate here, what I have previously stated—that those physicians who say that conception will not take place from the tenth or twelfth day after the cessation of the menses, till their re-appearance, assert what they ought to know is not so, as many of my readers, no doubt, know from their own experience; again: another proof of the absurdity of such a pretended discovery is, that females will often have their regular catamenial flow, for months after they have become pregnant. Any exercise that will disturb the embryo within twenty-four hours after conception, may be sufficient to prevent offspring. Dancing, and urinating immediately after, will often prevent. Riding a trotting horse, or any exercise, that will agitate the *ovum* before it is securely located, will certainly prevent. If these fail but once in five years, females cannot have large families. Among the other anti-fecundating remedies, are strong cathartics, all stimulating fluids, victuals that will promote thirst, bathing soon after coition, terminating the conjugal act before it reaches its ultimatum, will often prevent it; but, as I have stated in another place, previously, such a practice will produce seminal diseases and impotency in the male almost before they are aware of it. Prolonging the venereal act will also cause seminal disease and impotency.

A fine sponge, of an inch and a half or so in diameter, fastened by a silk string to withdraw it after absorbing the generating fluid, is ineffectual to prevent conception. The French *condoms,* or coverings worn by the male, are not only effectual in preventing conception, but are used to prevent contracting venereal diseases.

Three or four syringe-fulls of warm water energetically used immediately after coition, will sometimes prevent conception.

Neither the various medicines nor mechanical appliances, which are advertised as harmless, yet effectual, can be relied upon.

As neither of the preceding are positive and certain preventives to conception, and as persons including physicians from all parts of the country are applying to me every week by letter or in person, for a sure preventive for their wives, or some of their female patients, who cannot give birth to offspring without losing their own or the life of the child, I have after years of study, research and experiment, attained the great object which has been so unavailingly sought for heretofore, that is, for a man and his wife to fully enjoy sexual intercourse according to the laws of nature, and at the same time not beget offspring, or if they wish any, to limit the number according to their wishes or circumstances. It is used by the wife without inconvenience, and so secret that it cannot be known by the husband, and will last until the change in the wife's life, at about her fortieth year, renders its further use unnecessary, or in other words for a lifetime. It cannot cause the male or female the slightest injury, or interfere in the least with the fullest sexual enjoyment.

I have previously stated, that the French condoms will prevent conception, but the objection to them is, they are troublesome, liable to be torn in coition, prevent the full sexual enjoyment, as they do not allow that reciprocal warmth of the male and female genitals which is necessary, as it prevents their coming in contact except through the covering, which often chafes and irritates the female parts, and will if used for any great number of times, cause seminal weakness, debility and Impotency in the male.

Preventing conception and producing abortion are very different from each other; my plan does not allow the semen to enter the womb at all, which is the only possible way to prevent conception, as all of your family Physicians will inform you if asked. If the semen enters the womb, conception takes place, which cannot be got rid of without an abortion being produced, which all know endangers the life of the female, and leaves a lasting weakening disorganization of the sexual organs, sooner or later producing consumption or some general chronic disease, which proves fatal to the poor unfortunate; to prevent such suffering and calamities is the next great reason why I have consented to announce my remedy, as it must be plain to every one, that it will save an incalculable amount of suffering and crime, for from all the statistical information I can gain, there are thousands of abortions daily produced in this civilized, but wicked world. I am sorry to say, I am very often applied to by married and single persons, who offer me large fees, if I would produce abortion for them, and to save myself the trouble of any more such applications, I take this medium of announcing to the world, that I never have, and never will produce an abortion for any person, however large a fortune they might offer me for doing it.

The author is a Physician of a very extensive practice, and contrary to popular opinion, has the strongest sympathies for suffering humanity, and for a long time has made the physiological laws pertaining to the reproduction of our species and its attending conditions, a subject of close study. The nature of conception and the mode of subsequent developement, of the Fœtus, I have fully illustrated and explained in another part of this work. As I have before mentioned, the various plans which have been suggested by wise men, have either failed to answer the purpose, or were directly opposed to the full enjoyment of, or physically at variance with sexual union. Mankind will not adopt any remedy that will debar intercourse or interfere with the pleasure which nature has provided. *Nature is nature*, and all must admit, that it will not submit to being too much circumscribed. In fact you may as well try to suppress the sense of hunger and thirst by abstaining from food and drink, as to attempt to appease the equally imperative demands of sensual desires by some other method than the natural indulgence of the sexes. The apprehensions of accouchment, and the exhusting effects of nursing, to many, are too real. Many a fair women with spirit and vigorous health, will in a few years after becoming mothers, pine away into mere skeletons, with enough constant suffering to cause a living death.

Many married people of ardent natures are constrained to forego the pleasures of love, or else see their children inevitably afflicted with some incurable hereditary disease, which they would entail upon them.

The effectual prevention of conception, will greatly reduce the fearful crime of abortion, infanticide, and illegitimacy in many respectable families. Family discords, jealousy, and suspicions of infidelity, in a great measure emanate from excessive aptitude of conception, and dread of the dangers of confinement on the part of the female, as natures demands cannot be restrained When the husband finds that his dearest rights are invaded or denied at home, (many, very many such married men, I cure every month of venereal diseases,) he too often for the preservation of domestic peace, seeks unlawful pleasures abroad, his appreciation of home endearments diminishes, and his wife's suspicions of infidelity are too justly aroused, and discord, discontent, and very often a total rupture of the marriage relations ensue.

Many persons no doubt, will charge me with encouraging immorality and crime, which is the very reason of my declining heretofore, to offer this invention to the public; but being so incessantly importuned, and satisfied of the rectitude of my motives, flattered also, with the firm belief that on calm reflection, this will appear insignificant, and that I will be more than compensated by the immense amount of suffering and crime it will be the means of preventing; I present it to those whose welfare requires it. It certainly cannot be wrong to endeavor to legitimately promote our own happiness, or that of the common good of mankind. Is it a sin to prevent offspring, if it averts their being reared to swell the tide of ignorance, poverty, had moral degredation, that finds its end in the prisons? I certainly cannot bring myself to think so. It cannot be a sin to preserve the health of young mothers, thereby saving them the torture of broken constitutions, and an early grave! Certainly, to prevent the entailment upon our innocent offspring of an hereditary and incurable disease, or insanity cannot be! Those most interested may decide these interrogations for themselves.

Those who wish to avail themselves of our services, will remit our consultation fee of five dollars, giving us a general description of the female, and state whether she is suffering with a falling of the womb, whites (discharge), any other debility or not.

In directing letters to me, the number of my Post Office Box, 844 New York City, is all the direction that is necessary to write on the envelope, as they will reach me just as safe, as if my name was superscribed in full. N. B.—All letters containing money, should be registered.

MANY IMPORTANT TRUTHS IN REGARD TO THE HUMAN SYSTEM, WITH THE BEST MODE OF RETAINING SEXUAL VIGOR, TO EXTREME OLD AGE.

Coition.—Young persons often irretrievably injure themselves, by forcing the desire for coition. Nature must be your guide. Sexual commerce must not be prolonged or a fatal weakness may be the result. The virility of an old man will often be greatly increased and extended by marrying a young robust female. Young children should not be allowed to sleep with sickly or aged persons ; this advice should be strictly followed ; for we have too many diseased nurses in charge of our children imparting disease by the breath, or teaching them pernicious habits which may ruin them before they are discovered. Any excess in youth detracts ten-fold from future abilities in old age.

Child-bearing.—The usual period of pregnancy is nine months, though instances occasionally occur of its continuing ten or eleven months, and on the contrary, six or seven months only.

FRUITFUL MONTHS.—February, March, April and May, are supposed to be the most fruitful, or in other words, May, June, July and August, are most auspicious for conception.

TWINS.—Females may have twins—the offspring of different fathers. Colored and white children at the same birth have proved this. In fact, the mothers have acknowledged cohabiting with another person on the same day besides their husbands.

COLOR OF HAIR.—Fair, or red haired women are more ardent in their affections, and generally more fruitful.

MISCARRIAGES.—When a .female once miscarries, she will be always liable to miscarry when the same stage of pregnancy again occurs. Particular care should be taken at that period, by all females that have met with such an accident.

THE HYMEN.—The existence of the hymen in women is no certain evidence of virginity; and, *vice versa*, its absence is no proof of unchastity; though as a general rule, if virtuous, it is found at marriage. Illness and various accidents will rupture it.

EPILEPSY.—The only cure for uterine epilepsy is marriage.

VIOLATION.—Conception will take place, even when a female is violated by force.

HERMAPHRODITES.—There are no hermaphrodites,

PLATE 8.

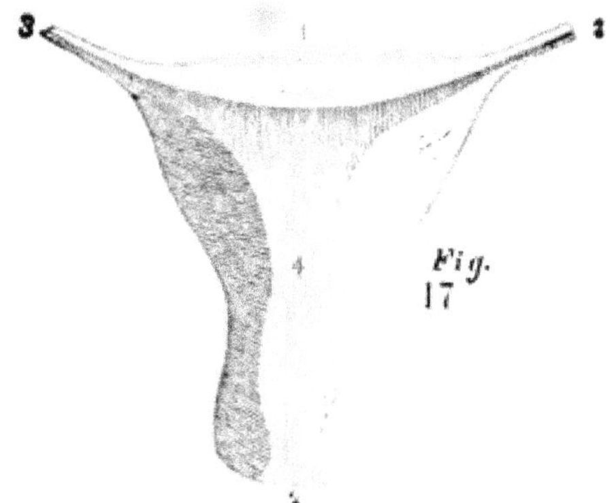

Fig. 18

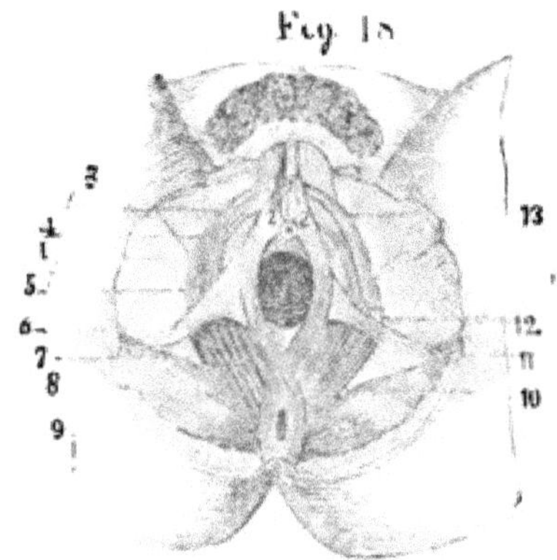

all mixed appearances which have been supposed to be such were caused by malformation.

Effects of Bad Temper.—The pregnant female should be protected from irritation, and govern her own temper, to ensure amiable offspring.

Signs of Pregnancy.—In addition to the usual known signs of pregnancy, an experienced observer will detect a glassy appearance of the eyes in the early stage of pregnancy.

Excesses—Total Abstemiousness.—Refrain—as I have previously stated the consequences—from youthful excesses. On the other hand, it has been maintained that total abstemiousness from sexual intercourse, would invigorate the mind, but facts prove otherwise. It is better to comply with nature than resist her altogether.

Suckling, etc.—A debilitated or sickly female should not suckle her infant. Yet, when not so, its own mother is its best nurse. Bottled porter is strengthening for the mother, but it should be used sparingly. If the mother will take exercise in the open air regularly while pregnant, it will benefit both herself and the child.

A few words on the Choice of a Partner.—The female should be from three to ten years younger than the male. If you are tall, choose a lady at least a head shorter than yourself. If you are of a lively disposition, your partner should be the reverse. A corpulent man should marry a spare women, and *vice versa*. A person of a dark complexion should be united with one that is light. In a word, married persons should almost be at direct antipodes to each other

I think it is directly in opposition to the laws of humanity for very sickly persons to get married This may seem a cruel interpretation, but I think it is more so, for parents to usher helpless infants into the world to lead out a miserable existence from hereditary diseases. By close observation for a considerable period of time, I find that in those couples where the age of the male is greater than the female, that a great majority of the children will be of the male sex, and when the female is the elder, they are of the female sex. Young ladies and gentlemen should test the correctness of this by examining into the cases of their friends and acquaintances. As they undoubtedly will find my researches correct, they should make choice of a partner in accordance with their future anticipations.

The most conclusive proof in favor of uniting opposites, is found in the evil consequences attending marriages among blood relations. In such persons there is generally a moral and physical resemblance, which is certain to entail suffering on their offspring. A glance at the degeneracy of the Royal families is the best of proofs that such unions should not be formed. In Spain the race has become puny, sickly, and imbecile. Scrofula—that most offensive of hereditary diseases—afflicts all the Bourbons, and the reigning families of Holland, Austria, England, and most royal families. It has become a generally known fact, that Queen Victoria has had a running ulcer between her shoulders for many years. It is with man, as with animals.

crossing the breed improves it. A writer says that the Persians, by adopting this crossing practice, have nearly obliterated the traces of their Mongolian origin. The most excruciating deaths of females at the time of delivery will be prevented, if persons will follow these and the foregoing directions in forming unions. A strong robust man should marry a woman with wide haunches, then the delivery of a robust child will be safe. A female, narrow through the haunches, should be united to a delicate, medium-sized man. A woman with narrow haunches always suffers extremely in giving birth to a child. A small one, therefore, will lessen it.

A modern physiologist truly says that a well-formed woman should have her head, shoulders and chest small and compact; arms and limbs relatively; her haunches apart; her hips elevated; her abdomen large, and her thighs voluminous. Hence, she should taper from the centre up and down. Whereas, in a well-formed man, the shoulders are more prominent than the hips. Great hollowness of the back, the pressing of the thighs against each other in walking, and the elevation of one hip above the other, are indications of the malformation of the pelvis.

If a female throws her feet much to the rear in walking, her knees are inclined inward. A woman that marches, rather than walks, has large hips and a well-developed pelvis. If she moves along trippingly on her tip-toes, a large calf and strong muscles are indicated. The foot lifted in a slovenly manner, so as to strike the heel against the back of the dress, is a sure sign of a small calf and narrow pelvis. A heavy walk, when there is but little spring on the toes, gives evidence of a weakness of limbs.

A man or woman having dark eyes and a consumptive tendency, should choose a blue-eyed partner. Confirmed consumptive persons should never marry. In partial or artificially produced consumptive cases, marriage is often very beneficial.

The most proper age for men to marry is between twenty-one and thirty; and for women, between eighteen and twenty-five years of age. The offspring of very early marriages is generally puny or consumptive: furthermore, their own health suffers, and their lives are shortened very materially. On the other hand, there is scarcely any freshness in the maiden of thirty; while the matron of that age, if her life has been a happy one, and her hymenial condition of not more than ten years standing, is scarcely in the hey-day of her charms. It is a well-known fact also, that bachelors *grow old* faster than married men.

Twins.—The extraordinary distension and attractiveness of the uterus, is probably the cause of the birth of one or more children at one time. Some writers think it is occasioned by the presence of several vesiculæ, ready to be detached from the ovaria, and consequently ripe for fecundation. They may be detected by motions in more than one part of the body at the same time.

Parturition.—When the child has become fully developed, the probability is, that the labor pains are produced by the reaction of the fibres of the uterus, thus causing a great distension, which makes it compulsory on the fœtus to evolve itself. After confinement the uterus immediately closes, and falls into a state of repose, from which it is not

PLATE 9.

Fig 19

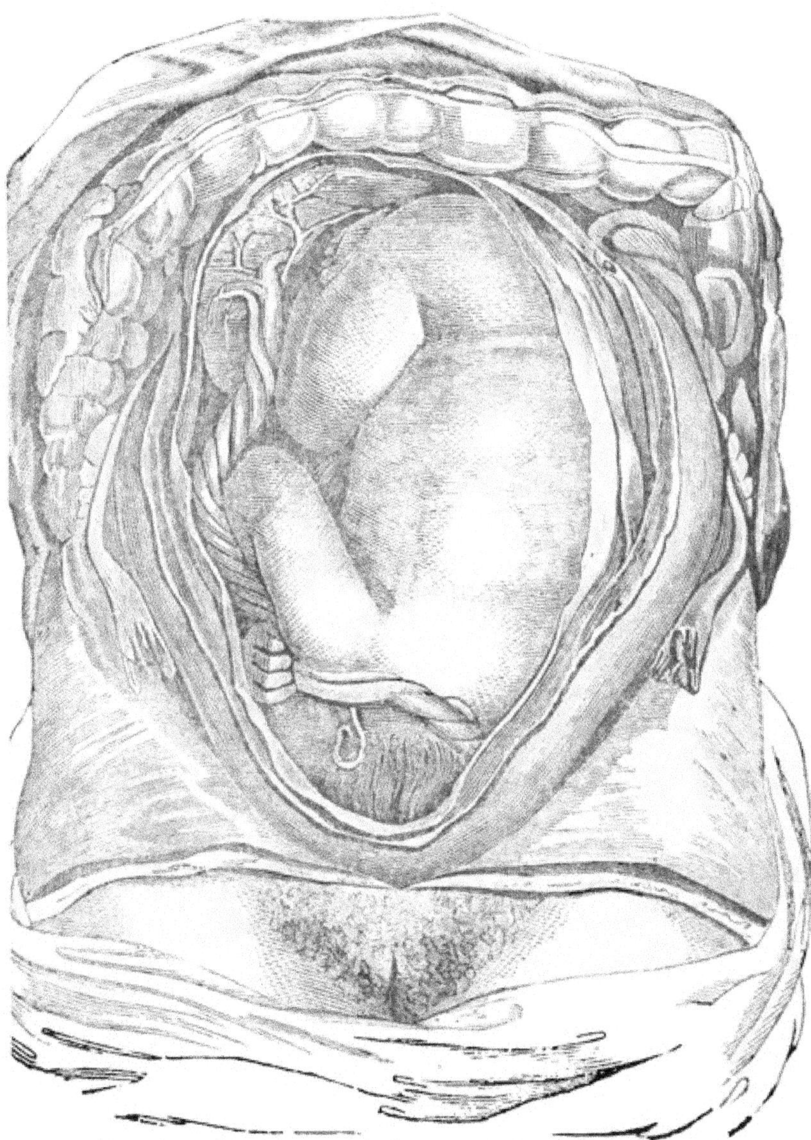

Fig. 19.—Represents a full grown fœtus in the womb, folded up and in its natural appearance and attitude, most beautifully represented and executed.

proper to disturb it for at least a month, as it requires that time to recover its natural situation.

FORMATION OF FŒTUS.—About the eighteenth day after intercourse, the ovum begins to assume a determinate structure.

NURSING.—A pregnant woman should not suckle her child, as it not only robs the fœtus, but injures the mother and child. The fœtus absorbs a portion of all the aliments the mother partakes of, therefore the necessity of pregnant women being careful of what they eat and drink. The milk taken by a healthy infant equals in weight about a third of the food taken by the mother.

TEETHING.—The small molar teeth appear between the ages of eighteen months and two years, and then the first dentition is complete. Children at this time of life will do better in the country, or on the sea shore.

RESEMBLANCE IN THE OFFSPRING OF THE PARENTS.—Notwithstanding all the speculative nonsense that has been written on this subject, there is nothing certain as to the particular faculties or appearance imparted to the offspring by the father or mother. There is only a general rule of resemblance. The mental organs of children, are not influenced by the frame of mind in which the parents—or either of them—may have been in at the period of reproduction. I think the fact of the semen not being created instantaneously—as it would have to be, for the offspring to be influenced by the state of mind of the parents at the time of coition—is a sufficient refutation of the theories promulgated by the *would-be great* physiologists. They and every one else, who pretend to any knowledge of this subject admit that the semen must remain a certain length of time in its receptacles for it to perfect its vitality. Each assist in modeling the embryo after their own form and likeness. The parent who is most energetic and excited at the time of sexual action, may impart the most distinct features of resemblance. But this combined resemblance is not imparted by one to one part, and the other to another, but an undefinable union governs the whole frame. The quality of one parent may preponderate, yet, that of the other, equally pervades the entire system. It is the same with color; the issue of a black and white, is well known to be of a uniform complexion. The children of aged parents—or where one is old and the other young—are generally delicate and spare of form, and very rarely well organized.

QUACKERY AND QUACKS.

Upon these subjects I have doubted the propriety and usefulness of an *exposé;* for it is customary—nay, to pretenders, an all-important theme, upon which to discourse learnedly—to stave off public odium and contempt from themselves, by assuring the public of their prodigious learning and skill, in tearing down the edifices of others to build a flimsy, tottering, untenable superstructure for the gigantic "*I, myself, the Doctor.*"

All humbugs depend for their success upon their being forced upon the world by dint of untiring, unceasing exertions to palm them off. I can estimate their success by learning the amount of exertion spent upon them.

As I have written a book or two, I shall not criticize a certain class of "Prodigies" in that line, lest I might accidentally fall into the awful vortex, and grow giddy from the elevated estimation of my own extraordinary "productions."

But I must regret what cannot be helped, that all large cities are infested with a set of self-styled "Professional Gentlemen," who care more for the pulses of the purse than for the welfare of their patients. That every man should look carefully to his own finances is right and proper, but a physician has other duties to attend to also.

It is all-important that Venereal patients have early, efficient and skillful treatment, as a simple case will assume a malignant form or undermine the constitution if neglected or maltreated.

In offering this plain practical work, I know those who follow its teachings, will give me credit for doing good, and of being candid.

Specifics.—Vended medicines, warranted to cure everything, are worse than no remedies; for in different diseases there are also different states of the system; and in different stages, even of the same disease, there is marked dissimilarity of action. In the inflammatory stage of Gonorrhœa, a warm bath is of great value in relieving pain, difficulty of micturation, and fever in the part; but no one of common sense would, therefore, advise a warm bath as the great remedy for clap. Yet this is what venders of patent medicines say, "Good for every thing."

Since I have been so serious upon a subject I purposed to treat lightly, I may be excused in giving a poetical effusion of a conscientious "Allopath of the Far West," which was shown me a few years ago by a western student, and which was regarded as a literary curiosity. I regret that I did not recollect the name of the author, for it would be particularly gratifying to me to pay genius due credit. The Poem is entitled—

THE HISTORY OF DOCTOR QUACKEY, JUNIOR.

His history—his fame must write,
Record a sacred truth,
Of Doctor Quackey's only son,
A great and famous youth.

A perfect tiger on fits was he,
A geater on a puke;
He kills the people all by bits (sitz).
He physics, good St Luke!

Although his practice was not large,
He managed it quite well;
Sent the pious up to heaven,
The wicked down to hell.

A *pillar* of the State was he,
And showed his mighty skill,
In filling up most speedily,
The grave-yard on the hill.

Hydrum-garum, herbum-scarum
A scholar sure was he;
Or never could such Latin words
Have uttered fast and free.

In fact, he was remarkable,
A man of genius pure,
With *Lobilly* and Highjohny,
Would all diseases cure.

To surgery he made pretence,
  Had a scalpel, spatula and fork,
Talked loud of London, Paris, Edinburg,
  Of Hospitals, New York.

Of cankers in the blood said much,
  Of heat and cold at war,
And swore by his huge saddle-bags,
  Red pepper was *the Cure!*

Devoutly did he thank his God,
  For him who invented Steam;
To parboil mortals of the sod,
  A hocus-pocus theme.

Don Quixote's balsam Sancho drank,
  Had ne'er such virtue sure,
Although it wrought a miracle,
  It left poor Sancho sore!

Gracious heavens!—help me conclude,
  In spite of tears and grief,
The history of that noble son,
  That charlatanic chief.

He died!—departed worth farewell!—
  Thy deeds, and prowess bold,
Hath reared high up to heaven,
  Fame's monument of gold.

He offered his life a sacrifice,
  To that dear, precious theme;
His all,—for to perpetuate,
  Lobelia,—Pepper,—Steam.

HISTORY OF VENEREAL DISEASES.

The term, Venereal Diseases, is applied to all those affections which are more or less, directly or indirectly, the consequences of sexual intercourse.

As early as two thousand four hundred years before the advent of our Saviour, the Jewish Lawgiver made special laws to prevent or cure diseases arising from sexual intercourse. Showing the evil to have been remarkably prevalent at that period to attract the attention of the government of that ancient people.

We will examine the thirteenth and fifteenth chapters of Leviticus, impartially, leaving the reader to form his own conclusions.

Leviticus, Chap. 13 : 2. When a man shall have in the skin of his flesh a rising, a scab, or bright spot, and it be in the skin of his flesh like the plague of leprosy, then he shall be brought unto Aaron the priest, or unto his sons the priests.

3. And the priest shall look on the plague in the skin of the flesh: and when the hair in the plague is turned white, and the plague in his sight be deeper than the skin of his flesh, it is a plague of leprosy: and the priest shall look on him, and pronouce him unclean.

8. And if the priest see, that behold, the scab spreadeth in the skin, then the priest shall pronounce him unclean: it is a leprosy.

14. But when raw flesh appeareth in him, he shall be unclean.

15. And the priest shall see the raw flesh and pronounce him unclean: for the raw flesh is unclean: it is a leprosy.

Whoever will read attentively these two chapters, will find that the diseases known as Leprosy and Syphilis, are considered by the ancients one disease. But I shall show that the preceding quotations indicate that the disease was Syphilis, for immediately after, a Bubo is described, which is the frequent consequence of Syphilis, and never of Leprosy.

18th verse. The flesh also, in which, even in the skin thereof, was a bile, and is healed.

19. And in place of the bile there be a white rising, or a bright spot, white, and somewhat reddish, and it be showed to the priest;

20. And if, when the priest seeth it, behold, it be in sight lower than the skin, and the hair thereof be turned white, it is a plague of leprosy broken out of the bile.

The following refer to the disease most known as Gonorrhœa, (clap,) and is of a different character from the former:

Leviticus, chap. 15 : 2. Speak unto the children of Israel, and say unto them, When any man hath a running issue out of his flesh, because of his issue he is unclean.

3. And this shall be his uncleanness in his issue: whether his flesh run with his issue, or his flesh be stopped from his issue, it is his uncleanness.

16. And if any man's seed of copulation go out from him. (This also proves that the involuntary loss of semen for excessive sexual indulgence was well known to Moses.)

19. And if a woman have an issue, and her issue in her flesh be blood, she shall be put apart seven days.

The ablution and abstinence enjoined by Moses must have been to prevent or cure affections liable to follow sexual intercourse at that time, and exhibits great sagacity and knowledge of human nature in the promulgator.

Solomon, 2950 years before Christ, was equally aware of the evil consequences of excessive com merce between the sexes, for in the fifth chapter, and third verse, we find Wisdom admonishes her son thus: For the lips of a strange woman drop as a honeycomb, and her mouth is smoother than oil. 4. Her end is bitter as wormwood, sharp as a two-edged sword. 5. Her feet go down to death. 18. Let thy fountain be blessed, and rejoice with the wife of thy youth. 19. Let her be as the loving hind and pleasant roe; let her breasts satisfy thee at all times; and be thou ravished with her love.

Later: St. Paul to the Corinthians, chapter 7, It is good for a man not to touch a woman; but if they cannot contain, let them marry: for it is better to marry than to burn.

"In the year 1245, Ubertius, governor of Carrara, died of a lingering disease of his private parts, occasioned by too much venery."

About the year 1423, a physician of Bologna, says, "that the retention of the poisonous matter lodged between the glans and prepuce, after a man has had to do with a foul woman, causes the part to become black, and the substance of the yard mortifies, and that Bubo occurs in the groin in consequence."

A physician of Rome, in 1527, first called it the Venereal Disease.

Not until 1782 was it demonstrated that Gonorrhœa and Syphilis were different diseases. The honor of this discovery is due to Benjamin Bell, an English surgeon.

In 1830, M. Ricord, a Paris surgeon, proved conclusively that a special cause (perfectly independent of the sexual organs) gave rise to, and occasioned the propagation of Syphilis, showing it to be as distinct a contagious disease by inoculation as is the small-pox.

Non-Virulent Affections will include Gonorrhœa, Leucorrhœa (whites), irritation and inflammation of the organs of generation, excoriations, etc.

Virulent Affections.—Syphilis is the result of a specific poison. First stage is local, second stage is, when it is absorbed into the system, and may be transmitted to the offspring, but not inoculable; third stage is still deeper seated, and affects the deeper seated parts, the bones, etc., and cannot be produced by inoculation, but it will be hereditary.

NON-VIRULENT DISEASES.

By the term non-virulent is meant those affections following (may exist without sexual intercourse) sexual intercourse, reproducing themselves, contagious by contact with mucous membranes, but will not produce the disease by inoculation in the skin.

Under this head, first we shall speak of those discharges common to both married and unmarried females.

It is estimated that four-fifths of the females of Paris have a constant discharge from the vagina. To attribute this to gonorrhœa would be monstrously absurd.

Even in this city female discharges are infinitely more common than many imagine, even in virgins; but they must not be attributed to contagion nor sexual intercourse, except in a small minority.

Yet when these discharges are acrid, they will frequently cause irritation and discharge from the male organ after coition, or occasion abrasion of the glans and prepuce, but which is not gonorrhœa. For example, it frequently occurs from a severe cold that the acrid mucous secretions excoriate the nose and lips, the tears, the eye-lids and cheeks, yet no one ever thinks for an instant that there ought to exist any suspicion of an immoral cause.

The mucous membrane lining the nose, eyes, mouth, vagina, penis or anus, is one and the same in its character. But if a discharge equally innocent, and which would cause no more irritation than the former, should exist in the male or female organs of generation, suspicion is at once aroused, but without a cause.

Then, from many of the discharges which females labor under, an irritation, excoriation or discharge may occur from coition, in the male organ, and not be a clap—proper—nor have been produced by previous intercourse.

DISCHARGES PECULIAR TO FEMALES.

Leucorrhœa is an excessive, and altered secretion of the mucus furnished by the membranes lining the vagina and uterus or womb, generally white or nearly colorless and transparent, without much odor, sometimes gluey or resembling pus, (matter,) or may be yellow, green or slightly bloody, either very thin or quite thick, and may exist for a few days or many years.

This is an affection among females which is exceedingly common, and which frequently cause a disease in the urethra of the male—even in the staid, sober, pious, and strictly moral man too; but there are those whose piety is questionable, who try to shield themselves behind false assertions, and flagrant improbabilities.

The cases are so rare in which persons are happily deceived, that I shall relate one: M——, a married man, of family; like a certain Congress man, whose sympathy induced him to offer a peti tion in Congress for the Relief of Widows and Orphans.

He said, "Mr. Speaker, (holding a paper in one hand, the other being thrust into his breeches pocket,) I hold in my hand a petition for the relief of Widows and Orphans." A waggish member asked the orator, "In which hand?" The difference between the latter and the former gentleman was, that the first had sympathized with a widow only, and that his pungent convictions were that he was not the first who had also sympathized with the "Lone Widow."

But the result proved that his fright was occasioned by irritation of the urethra from a leucorrhœa, which terminated the menses. However, to punish him for his waywardness, I let him repent for his sins of commission, by not undeceiving him.

I must necessarily be brief in the little I have to say upon the treatment of leucorrhœa in either sex, or rather in the female, and the irritation caused by it in the male. Cleanliness is an important item here. In the male all these irritations as stated above, will yield in a day or two with the appropriate treatment.

But in the opposite sex, leucorrhœa may depend upon debility or upon the robustness of an inactive life so common among people in easy circumstances, or upon excessive venery or copulation. The bowels should be kept regular with Turkey rhubarb, and frequent injections into the vagina of warm water should be used. Moderate and out-of-door exercise, and good, but not stimulating, diet are also proper. In the robust, a recent affection will be relieved by a cleanliness of the parts, frequent but gentle physic and exercise and low diet. Of course, coition must be prohibited for a time at least.

I was consulted by a married lady who had suffered extremely in body, as well as disappointment from sterility for six years, in consequence of an inveterate leucorrhœa (whites), which to her was almost death, as her whole heart was full of the desire and hope to have children. After three weeks' treatment—simple treatment—she suspected that the object of her sanguine hopes was attained, as the monthly period had passed without the usual menstruation.

Time revealed the fact that the "consummation most devoutly to be wished," was no longer an uncertainty.

I am satisfied that a large number of married women are barren from this cause.

OF AMENORRHŒA, OR THE ABSENCE OF MENSTRUATION.

First.—Amenorrhœa of Rentention, in which the function is not performed at the natural time, or at all, of which there are three varieties.

1st,—Non-appearance of the menses in consequence of congenital deficiency, or malformation, or strictural disease of the genital organs.

2nd,—Want of menstruation, where independently of deficiency or malformation, there is either a show and partial development, or an entire absence of puberty.

3rd,—When the function ceases after puberty.

Second.—Suppression of menstruation after it has been established for a time, independent of pregnancy or lactation, including two varieties.

1st,—Recent and acute suppression.

2nd,—Chronic suppression.

In a work like this it will the better serve the end for which it was written, to give a succinct description of those of more frequent occurrence, leaving the rare and extraordinary to our work designed for the professional reader alone.

Suppression of menstruation after puberty may depend upon too great fullness of blood, robustness of habit, seldom occurring in those young females who reside in crowded towns or manufacturing places, but in young ladies of the country who live more naturally, and exercise in the open air. It more usually occurs in the delicate, irritable and hysterical.

Amenorrhœa in females of full habit and costive tendency, is attended with full, heavy oppressive feeling in the head, back and loins, with either great heat of the hands and feet, or coldness, with sudden flashes of heat; drowsiness, dizziness and indisposition to mental or physical exercise.

The appropriate treatment, will be regular habits, low diet, and frequent physic.

The compound Aloetic Pills are an excellent cathartic. Dose, two to five, taken at bed time, twice or thrice a week.

Amenorrhœa from a delicate, irritable or hysterical state of the system, is diametrically opposite to the former condition of system; the former being too full of the vitalizing fluid, the latter being not full enough.

In a young or middle aged woman, fleshy, full of blood, high color, the suppression of the uterine function, will be followed by congestion or inflammation. While in a delicate woman, thin, spare, of sallow complexion, extremely nervous, the result of suppression would be irritation, attended with spasms and severe pain, with intervals of ease and quiet. In the former, violent hysterics often follow an attack. In the latter, the symptoms—as pain in the head, back, loins, limbs and abdomen, yet it is of a neuralgic or nervous character. The pain attacks first one and then another organ, frequently shifting.

If treatment is resorted to, such as a mustard poultice, or a stimulating and anodyne formentation, the pain is quickly removed from the womb to the head, from the head to the chest, or heart, or bowels. The patient is subject to fits of hysterics and fainting.

In treating the debilitated female for the suppression of the function of menstruation, the two grand objects are to increase the strength and improve the general health, and to enrich the blood. This is best accomplished by generous diet, such as is easy of digestion, keeping the bowels regular by means of the compound Aloe pills, and the internal use of iron. The muriated Tincture of Iron taken two hours after meals in half a wine glass of water in the dose of three drops, increasing one drop per dose at the end of every three days. An occasional mustard hip bath, cool sponging followed immediately by friction of the surface, to get up a glow, are powerful auxiliaries.

Of direct applications to the womb, electricity is almost the only means to be used.

In some of the cases, where, after the condition of the stomach and bowels were healthy, the amenorrhœa continued with slight paleness and weakness, electric shocks passed through the loins quickly induced menstruation. In others, its frequent repetition led to a similar result; and instances were not wanting where a shock suddenly produced the flow. Electricity is a powerful agent and must be used with great caution. If pregnancy be suspected to exist, however strenuously denied by the patient, it ought not to be employed. Doctor Ashwell once ordered it, he says, quite ignorantly, where the suppression depended upon pregnancy cautiously concealed, and abortion occurred within an hour.

OF MEDICINES WHICH EXERT AN INFLUENCE IN RESTORING MENSTRUATION

Mercury is a valuable and powerful remedy, but it should only be administered by a competent medical adviser.

Iron, I have already spoken of.

Ergot of Rye—Sacale Cornutum.—This medicine has wonderful influence upon the pregnant uterus, by producing contraction of that organ, hence it is not safe in the hands of persons who wish to secure themselves from the opprobrium of society, to destroy the evidences of criminal indulgences.

Iodine, Madder, Rue, Savine, Seneka-root, Nitre, Digitalis, Gamboge, Serpentaria, Wormwood, Musk,

Myrrh, Castor, Valerian, Lavender, etc., etc.; but most of these are simply adjuvants.

Aloes is the most valuable emenagogue cathartic, producing the most salutary effect upon the uterus through the bowels. This is so serious and so common a difficulty, and the results, if neglected, so melancholy, that I most earnestly advise all such sufferers to obtain early and good medical aid.

I have seen many—very many, beautiful, talented, and accomplished young ladies go down to premature graves by the cessation of this all important function of the uterine system.

GONORRHŒA IN THE MALE.

In France it is commonly known under the name of *chaude pisse;* in England and in this country, it is called clap, from the French *clapier*, which signifies a filthy abscess.

The circumstances which favor its contraction, are a large meatus in the penis, and the fact that, during erection and previous to the ejection of the semen, the orifice of the urethra has a great tendency to open, thereby favoring the introduction of the gonorrhœal matter. This undoubtedly occurs previous to emission, otherwise the disease would be more common than it is, if it was not forced out of the urethra by the ejaculated semen.

The symptoms of Gonorrhœa consist in, at first, a sensation between pain and a pleasurable feeling near the end of the penis, just within the urethra, which frequently excites erections. Soon after, however, follows an increase of pain, heat, redness of the meatus, scalding on making water, and a thick yellow or greenish-yellow, discharge from the

PLATE 10

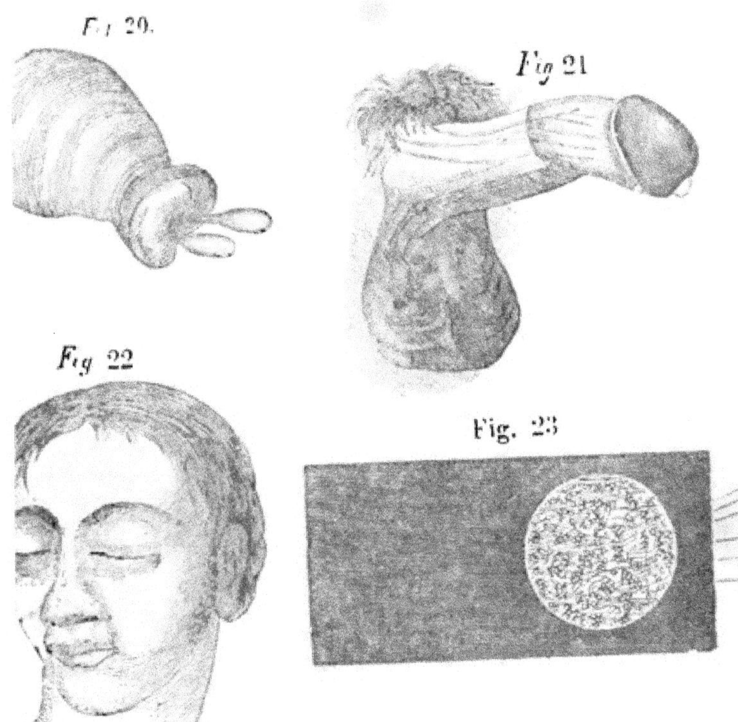

urethra. Sometimes, there is great difficulty of passing urine, with excessive inflammation of the end of the penis and prepuce.

These symptoms are soon followed in many cases by Chordee, which depends upon the loss of elasticity of the tissues which surround the urethra, and which during erections do not yield as in health, so that the penis is drawn downwards or bent to either side, exciting great suffering in those whose organs increase much in size during erection.

In severe cases, cramp, pains in the testicle, a feeling of tightness in the groin, pricking in the perineum below the testicles, and frequent desire to pass water, add to the severity of the case.

These symptoms increase as the disease travels down the canal.

These are the more urgent symptoms of the acute stage.

Chronic Stage—Gleet.—In consequence of no treatment, or that which is inefficient, the acute stage degenerates into the chronic, which is known by a marked mitigation of all the symptoms of the acute form. The discharge may become thin and watery, or be thick and less copious. In other cases, no running is seen during the day, the lips of the meatus being merely glued together; the slight discharge upon the linen may only leave a stain like that of gum. It sometimes occurs that the only symptom, and one which is apt to frighten fidgety (nervous) persons, is a discharge of shreds of mucous which resembles false membrane, or bits of vermicelli.

The Chordee may continue a considerable time after the other symptoms have ceased.

Occasionally, when the Gonorrhœa is nearly

well, nocturnal pollutions occur, and not only occasion great suffering, but excite a renewal of all the former symptoms.

Difficulty of making water (mictarition) may occur in this stage, which is caused by a spasmodic stricture, or inflammation in the canal, from taking cold. Or an opposite state may supervene. The urethra loses, in part, its power of contracting, so that when the bladder has emptied itself, a small quantity of urine remains in the canal, which it is not able to evacuate, and as soon as the penis hangs in a dependent state it dribbles away, staining the clothes, and is no inconsiderable annoyance to the patient.

As I treat fully in another place upon Impotency, Seminal Weakness, Masturbation, and Sterility, I will simply state that chronic gonorrhœa, or gleet, will almost invariably cause impotence and sexual debility. The patient should not forget this important truth, that is, that a private disease (venereal or gonorrhœa) never dries up, or wears out. It will certainly continue or end by producing another disease, if not radically cured.

I always have some patients—which, unfortunately for themselves and wives—can bear witness to this fact. No person, therefore, who has cohabited with any female—other than his wife—should even think of sexual commerce with her, for at least two weeks after the other coition. I cure innocent wives weekly, whose husbands have given them the disease, from not being aware of the necessity of refraining from having to do with them the length of time above stated. I have cured patients who had not cohabited within fifty days, but at the expiration of that time cohabited with their wives, and the next day the disease made its appearance, although they had felt, and the penis had looked perfectly well for the whole time. I have so many cases of disease that does not develop itself till a day or two after cohabiting with their wives, that I fearlessly assert as an indisputable fact, that nine out of every ten females will contract a disease from a man who has cohabited with a diseased woman, though the disease may not be developed for a day or two after the healthy connection. As a proof that his wife is pure, no private disease generally develops itself within five or six days after the connection. A chancre (primary syphilis) may show itself—in consequence of an abrasion of the skin of the penis in the act of coition—within twenty-four hours, but under no other circumstances will it do so.

In consequence of the immense number of half-cured, constitutional and complicated cases, constantly being revealed to me, I think the best advice a person can follow who has been inoculated, is to open the bowels, adopt a spare diet, keep the parts clean with cold water, and apply immediately to a competent physician, but, above all, never use the quack or advertised remedies. The hands, and your clothes, must be kept clean and free from any discharge, or your eyes, nose, or anus will be in danger of inoculation, as the smallest possible quantity is sufficient to create a disease.

Chordee.—By following the advice just directed, Chordee will seldom succeed, but when it does, immediate attention will be required. As erections are the main cause of suffering, they may be avoided or checked by dispensing with female society, lascivious thoughts, stimulating and late meals, feather-beds, and much bed-clothing.

When erections occur use cold bathing to the penis, and put the feet on the cold floor. An emulsion, containing 5 grains of camphor, and an eighth of a grain of morphine, (sulphate of morphine,) should be taken at bed-time. The camphor can be pulverized by adding a drop or two of alcohol; sweetened water, syrup, or gum-arabic water, can be used as a vehicle for the medicine, which must be continued as long as the chordee is distressing.

Retention of Water.—This will sometimes happen, and will be relieved by hot mucilaginous teas drank freely, while a hot poultice is applied to the lower part of the bowels. Should this not succeed, the water must be drawn with the catheter.

Inflammation of the Neck of the Bladder.—Which is indicated by a frequent and urgent call to make water, attended with pain on passing the last drops, and which may be mixed with blood. Mucilaginous drinks, low diet, and an enema, containing 20 to 30 drops of laudanum, must be thrown into the bowels twice-a-day. If severe, even more active means should be used, for which the advice of a physician will be necessary.

Buboes.—When Buboes occur in the inflammatory stage of Clap, they are only sympathetic, and require a hot poultice, or possibly a few leeches to be applied, and the bowels to be kept open. Sometimes, however, this class of buboes—especially if the patient is of a scrofulous diathesis—are of the most obstinate kind.

SWELLED TESTICLE.

Generally, during the continuance of Gonorrhœa, if the patient is at all observant of his symptoms, he will feel pain in the perineum, accompanied by a dull, heavy, aching sensation in the groin and along the course of the chord of the testicle, and finally a settled pain in the upper and back part of the scrotum, (bag,) so that the hardness and swelling can be traced with the thumb and finger for a considerable distance along the cord. If the patient frets the organ by walking, or takes cold, the swelling is rapidly increased, the suffering on the slightest motion is excruciating. Often during sleep the suffering is suddenly rendered almost insupportable by the occurrence of nocturnal emissions. The semen may be accompanied with blood, which sometimes gives temporary relief.

When the inflammation runs high, there will be fever, dry skin, furred tongue, hard quick pulse, pain in the testicle and belly (abdomen), sometimes constipation, accompanied with vomiting. But notwithstanding these severe symptoms, patients seldom die of this affectiom.

The causes which excite this disease are cold, falling of mumps, fatigue, damp weather, sexual intercourse, and particularly are these causes more apparent, after gonorrhœa, which leaves the organs of generation very susceptible to slight influences. The inflammation in the urethra may travel to the testicle directly.

Swelled testicles seldom occurs during the inflammatory stage of gonorrhœa, therefore I advice early and powerful means, which, instead of creating the disease of this sensitive organ, cures the primary condition of the uretha, thereby precluding the possibility of its occurrence. The cord leading into the abdomen from the upper part of the testicle may also swell and be painful.

To prevent the difficulty, Gonorrhœa should be cured as early as possible, and when there has been uneasiness in the organ, a supensary bandage to hold up the part should be worn. When the difficulty has taken place, staying at home is of but little use unless the recumbent posture is perseveringly maintained, and the testicle supported, as its own weight will help to aggravate the inflammation and augment the pain. The diet must be low, and all spirituous liquors, beer, etc., must not be allowed. The gonorrhœa—which is the usual cause of swelled testicle—will re-appear, or increase, as the swelling decreases, for it is a singular fact, that the discharge may cease during the whole time of the swelling of the testicle.

GLEET.

A Gonorrhœa, unaccompanied with pain in urinating, is properly a chronic one, and a gleet is a slight discharge just before urinating, or may be only on rising in the morning. It will sometimes be so slight, that the lips of the urethra or meatus are glued or stuck together. This is often the case, when a stricture has formed. Chordee may continue for a long time after the original disease is entirely eradicated, unless it is also removed by the proper treatment. It is a very usual occurrence for me to cure patients who have been under treatment for even twenty years, yet in nineteen cases out of every twenty, I can cure them in two days, and often one.

This concludes what I have to say upon Gonorrhœa in the male, except a few words upon Gonorrhœal Rheumatism, and Sore Eyes.

It is a most remarkable fact—and one, the solution of which is far from being clear, or satisfactory—that a person who has been affected with Gonorrhœa, for only a few weeks even, will often suffer from *Rheumatism* and *Sore Eyes*. And what is more remarkable than that which I have just stated, is, that the only successful treatment for this sort of constitutional or sympathetic disease, is the same, as the one I adopt in curing *constitutional Syphilis*. The rheumatic affection, will very often continue for a long time after the local one has been cured. When the eyes are inoculated with the gonorrhœal virus, the rapidity and violence of the disease will destroy them in twenty-four hours even, if an heroically prompt treatment is not at once adopted.

GONORRHŒA IN THE FEMALE.

This affection is scarcely different—except in a few particulars—in its effects and symptoms in the female than in the male. As the female organs of generation are directly opposite in construction in the sense that they are calculated to receive the male organ, the capacity alone, renders the symptoms and suffering from gonorrhœa in the female much less urgent and less severe.

In fact, a female may have a discharge and even a gonorrhœa, and know nothing of its existence except from the stains found upon her linen.

The usual symptoms, are heat, uneasiness, discharge, and sometimes, but not always, smarting while (micturating) making water.

As the disease is purely local, the general system is but little affected. The spirits and digestion may flag for a time, and the internal lips may be swollen, red and painful; but no difficulty will attend passing the water. On opening the inner lips, (nymphæ,) a thick yellow, or greenish-yellow matter will be observed issuing from the parts. When, however, the disease is situated in the urethra, there will be some pain in passing urine, but not so severe as in the opposite sex, for the canal is short and much larger.

When the gonorrhœa is deep in the vagina, or in the uterus, (womb,) the cure is not so rapid, (in fact, is sometimes almost incurable,) as when less deep, or in the urethra. Frequent injections of water, to thoroughly cleanse the parts, is of the utmost importance. The female syringe should be of medium size, and bent at an obtuse angle, to enable the patient to use it herself. The injection should be gently thrown into the vagina; the beak of the instrument having been passed two or three inches, while the patient is in a bath, or on the edge of a seat. To retain it a few minutes, it will be necessary to lay upon the back with the hips slightly elevated.

When the affection is in the outer portion of the vagina, dry lint, or soft linen cloth may be passed into the canal after injections have been used, to keep the irritated and inflamed surfaces apart, which will have a salutary effect.

In deep-seated inflammation of the inner lips, or nymphæ, suppuration will sometimes take place. The matter which forms, will produce a large and painful swelling of the part.

PLATE II.

Fig. 24.

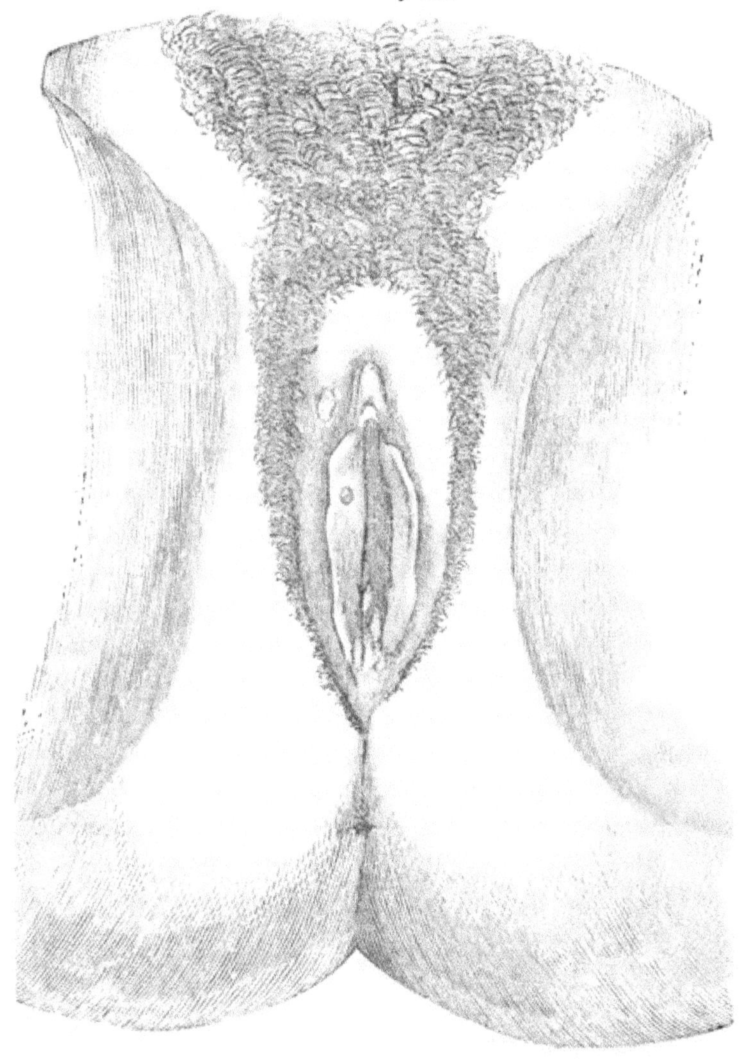

EXCORIATION OF THE GENITALS.

Under this head I shall treat of all simple abrasions, fretting or chafing of the organs of generation in both sexes.

Excoriation, and little water, or other pimples are very common in some persons. They are caused by gonorrhœa or syphilis, not having been thoroughly eradicated—an irritation in the urethra—or by excessive sexual intercourse. It is more common in newly married people, and they should regard it as a kind warning against a too free indulgence in the pleasures of the honey-moon, by that prudent old matron, Dame Nature. For those who heed not, shall suffer the penalty of her stern laws.

STRICTURE.

Stricture is a common result of gleet, or a gonorrhœa, long continued.

A stricture is a lessening of the natural size of the urethra.

Sir Astley Cooper's classification of strictures has been followed by most of the eminent surgical writers since his day, viz.:—the *Permanent*, the *Spasmodic*, and the *Inflammatory*.

OF SPASMODIC STRICTURE.

Sir Benjamin Brodie has so graphically and correctly described this form of the complaint, that I shall quote his own words: "A man who is otherwise healthy, voids his urine one day in a full stream. On the following day, perhaps, he is exposed to cold and damp; or he dines out, and forgets, amid the company of his friends, the quantity of champaigne, or punch, or other liquor, containing a combination of alcohol with a vegetable acid, which he drinks. On the next morning he finds himself unable to void his urine. If you send him to bed, apply warmth, and give him Dover's powder; it is not improbable that in the course of a few hours the urine will begin to flow. After the lapse of a few more hours you give him a draught of infusion of senna and sulphate of magnesia, and when this has acted on the bowels, he makes water in a full stream."

As the affection is purely a local one, general treatment cannot be relied on too much, as no time should be lost. The emergency in such cases are great, and require prompt and efficient action.

Under all such circumstances, I advise that a physician should be called if practicable. The object and aim of these pages is not to discourage or depreciate the services of physicians, but, on the contrary, to elevate the standard of the profession in the estimation of the public, and particularly with that unfortunate class who may have fallen into the greedy grasp of medical vultures, whose scent for gold is prodigious and unerring.

INFLAMMATORY STRICTURE.

This form of stricture arises usually in the acute stage of Gonorrhœa, or from the improper violence in the use of the catheter or bougie. The symptoms are fever, pain, and inability to make water. Should a few drops escape, the torture is intense.

Leeches to the groin may be necessary, but

usually a free use of the infusion of salts and senna, and a warm, sitting bath, will give relief in a few hours. After the water flows, barley water, gum-arabic water, flax-seed tea, are all that will be required. Instruments must be avoided.

Young men are almost the exclusive subjects of this form of stricture.

The frequency of the attack will depend very much upon the irritable condition of the system. The difficulty is a spasm of the urethra near the bladder, and sometimes the suddenness of the attack and the severity of the symptoms are extremely alarming to the poor sufferer and his surrounding friends. The simple form which Sir Benjamin has so beautifully described, and which we have just quoted, may become greatly aggravated and require further treatment.

The desire to make water occurs without the ability to accomplish it. After several unsuccessful attempts, the patient becomes alarmed—he now suffers great distress in the region of the bladder, and not unfrequently pain is felt in the end of the penis. The efforts to make water are now constant and beyond his control, the muscles of the abdomen contract violently, and the restlessness and contortions of the body from pain, are agonizing to behold. The face is flushed, the tongue coated with a white fur, the pulse hard and bounding, the skin hot, and the countenance of the patient is indicative of the most excruciating suffering

PERMANENT STRICTURES.

In the present form under consideration, it will be borne in mind, is found that unpleasant difficulty which is generally understood by the common term Stricture.

It is usually the result of chronic gonorrhœa, but may occur from any other inflammation, or from injury. It is quite common in cases of seminal weakness.

This form of stricture may depend upon an alteration of the surface of the mucous membrane, of the narrow canal or urethra, or upon a thickening of that membrane: in either case there exists a mechanical obstacle to the passage of the urine.

It is unnecessary to enumerate, in a popular work like this, the various alterations of stricture, from ulceration, fungus or warty excrescences, etc., as it will be of but little practical value to the unprofessional reader.

The situation of Stricture is more commonly met with in the membranous and bulbous portions of the canal, near the neck of the bladder; but any part of the urethra may be their seat.

The accompanying cuts will give a correct idea of Strictures, although they are generally more simple.

SYMPTOMS OF STRICTURE.

Persons may have Stricture for a long time if they are inattentive, or ignorant of the earlier symptoms of its incipient stage. The water passes with less freedom than natural, and finally the stream becomes smaller and smaller, and spirts out in several spiral or cork-screw streams, soiling the clothes or dropping upon the feet of the patient drop by drop, requiring time and patience to evacuate the bladder. Sometimes persons have no

PLATE 12.

Fig. 25

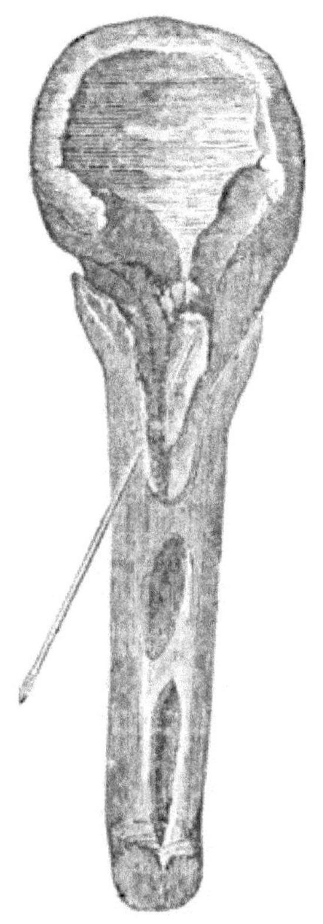

control over their hydraulic powers, there being an almost constant dribbling away of water.

In chronic bad cases, the prostate gland, neck of the bladder, ureters and kidneys, become inflamed, ulcerated or gangrened, gravel and stone will be formed, and the rest of his short life be almost intolerable. The prostate gland will suppurate, and it has long been observed, (and I have cured many such cases,) in bad cases of stricture, the urethra becomes so impervious during erection, that the sperm cannot be ejaculated, but escapes with the urine when the penis is flaccid.

Gleet may be the only symptom of stricture; therefore it often leads one to the cause of a long continued running, which may assume all the appearance, at times, of genuine clap. Sometimes, nothing but little shreds of mucous is discharged resembling, by the magnifying powers of a hypochondriacal imagination, vermicelli or worms.

This bug-bear is held up in silly and injurious books, to frighten patients to apply to the quacks, who practices copious depletion of the *Purse;* and the practice is wonderfully efficacious, allaying all apprehension of the devouring propensities of these animals.

M. Ricord, a French Surgeon of great eminence on Venereal and Genito-Urinary Diseases, says, "I am well aware, that strictures are often more quickly cured in proportion as they are early treated." But I have permanently relieved strictures which had existed for ten to twenty years, in nearly as many minutes.

The practice of allowing instruments to remain in the bladder during the night, as well as their frequent use, are not only unnecessary, but exceedingly mischievous. I am often consulted by persons who have pursued such a pernicious course of treatment, who labor under all the distressing symptoms of irritation of the neck of the bladder, or from chronic inflammation of the prostate gland and bladder.

I shall not go into a discussion of the pathological results of the treatment of stricture by the use of the bougie.

It has been my design from the first time I put my pen to paper in view of writing a practical treaties, to avoid enticing indulgence in theories and vain speculation.

The genius of the age, is practical, and especially so in this *Yankee* part of Christendom, wnen all the elements and nature are whipped into the traces, and guided by the practical hand in the race of wind, steam and lightning. Truly, the awful thunder-bolts of heaven have been tamed, and are being sent at will, by the commands of the humble creature, man.

DISEASE OF THE PROSTATE GLAND.

SYMPTOMS.—Those who labor under disease of the prostate gland, experience uncomfortable or painful sensations in the affected parts, or in the space between the scrotum and anus, or near the margin of the latter, which vary in different individuals, and the extent of the disease. The symptoms are increased during the evacuation of the bowels, or urine, also after exertion on foot or horseback, or vehicle riding. The sufferer experiences frequent and urgent desire to urinate, which is often felt so suddenly that it is irresistible; he will feel more or less pain near the neck of the bladder, particularly

PLATE 13.

Fig. 26

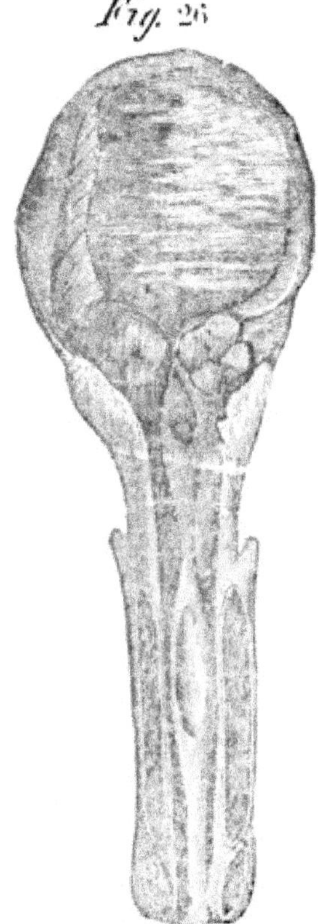

at the commencement and termination of the flow of urine. He will feel as though there was more urine coming out, but he will not see it till after the penis becomes pendant.

Such cases are extremely common in aged men, or those who have abused the sexual function. The patients often suffer from Stricture, or sometimes almost obliteration of the urethra, within an inch or so of the prostate. I have cured many patients of this complaint.

Chronic Inflammation of the Neck of the Bladder.—It is evident that there is a close affinity between diseases of the prostatic portion of the urethra and those of the neck of the bladder. Inflammation of the urinary canal, may not only extend along its whole extent from the external orifice, but to the neck of the bladder, prostate gland, orifices of the seminal ducts, seminal receptacles, deferent ducts, and testicles, but also along the inner or mucous surface of the bladder, hence along the ureters, and from these to the kidneys. There are some superficially informed physicians who consider Gonorrhœa a very trifling disease, but their ignorance can only estimate its commonest symptoms. This distressing complaint will often last a lifetime, as I have before stated. It often lays the foundation of many incurable diseases, not only of the genito-urinary organs, but likewise of those in the head, chest, and abdomen. Purulent discharges are almost always present in diseases of the prostate and bladder. A severe cold, excessive use of spirituous liquors, or too frequent coition, will probably be the first warning the patient will have of his unfortunate situation.

I will here warn persons against the habit of retaining the urine in the bladder after nature has exhibited, by the desire, the fact that her calls should be attended to, and the bladder should be evacuated at once.

IRRITATION OF THE TESTES.—When the testicles are excited or irritated, they secrete a much greater quantity of semen than in an ordinary condition; and this sperm is more watery and much less elaborated, and remains a shorter time in its receptacles, in which its watery parts are absorbed, while it is much more promptly evacuated, because the seminal receptacles are more sensitive to its excitement, and therefore more readily contract. This causes impotency in many men, as the penis no sooner touches the female genitals than the semen oozes out, the erection of course goes down, and will probably remain so for hours. And then, when again erected, they are subjected to the same disappointment and mortification. Hundreds of young and old men, newly married, have been placed in this horrible situation for weeks without accomplishing the marital duties, and then—notwithstanding their persevering efforts—have been obliged to apply to me to cure them. I have at least one or two such patients, always under my care. It is not necessary for me to ever hint at the situation of the poor female, who has been tantalized for such a great length of time, for I cannot think there is a man heartless enough to marry if he is aware of his unfortunate situation. The object of this work is to guard and enlighten both sexes.

It is evident from the foregoing statements, that the kidneys, ureters, bladder, vesiculae seminales

PLATE 14.

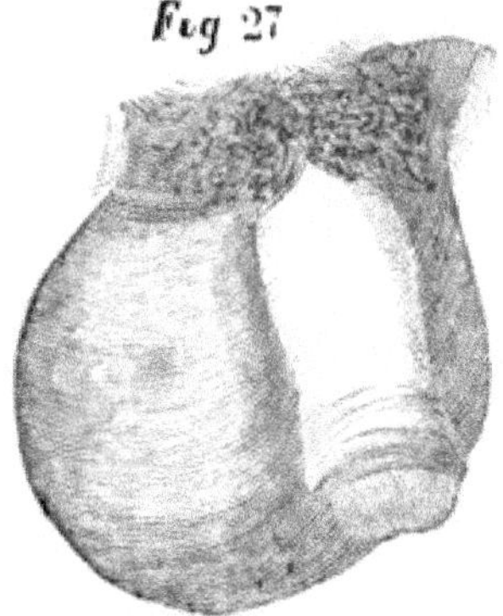

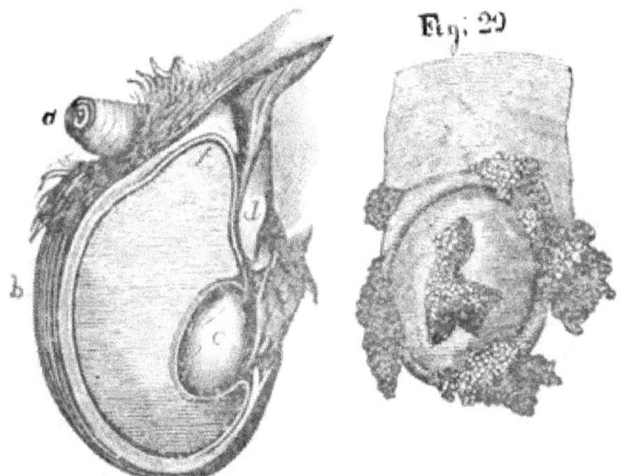

Fig. 28. The Scrotum distended to its utmost extent, and the position of the fluid. The Penis is almost always more or less drawn up, and in severe cases it appears drawn up so as scarcely to be perceptible.

and testicles, are under the influence of the same causes, which produce the same effect.

Varicocele and Hydrocele.—Arise from masturbation, venereal excesses, diseases, etc. We often see venous enlargements of the spermatic cord from all such causes. *Circocele* and Varicocele generally occur from the adult to old age, from these causes.

Diseases of the Kidneys and Ureters.—The same excesses will disease these organs.

Infiltration of Urine, caused by stricture or inflammation, will soon destroy the *Testes*, *Scrotum*, and *Penis*, and, of course, terminate in death.

Hæmorrhoidal and other circum-anal diseases arise from the same causes.

Irritation of the Vagina and Uterus frequently causes *Sterility*, *Schirrus*, and *Cancer* occasionally. Venereal excesses of whatever kind will produce *Uterine Hæmorrhage*.

Vegetations.—The urethra and vaginal discharge is so acrid in many cases as to excoriate the labia, and to give rise to excrescences or vegetations, which sometimes is succeeded by ulceration or sloughing.

Phymosis and Paraphymosis.—When the prepuce is inflamed and swollen, so that it cannot be drawn behind the glans penis, the disease is termed Phymosis; and when this part, after it has been drawn behind the glans, and cannot be drawn over it, the disease is called Paraphymosis In either case, there is danger of inflammation, sloughing (mortification), and more or less destruction of the penis. In cases of phymosis, when the glans cannot be uncovered, the internal membrane of the prepuce becomes inflamed from the accumulation of the natural secretion, or from the acrid matter of chancres, or urethritis, and an artificial opening may be caused by ulceration.

There is natural or congenital phymosis, which impedes the urine; also in adults, compressing the glans, (head of penis,) during erection, thereby preventing proper sexual commerce. In case the contraction of the prepuce is so small as to obstruct the evacuation of urine, after it has escaped from the urethra, irritation, inflammation, or sloughing, will ensue. Cases of rapid sloughing, and destruction of the penis in aged and other persons, are quite frequent. All persons thus affected, should immediately apply for proper treatment. I relieve many such persons, in a simple manner, without leaving any traces of their former trouble very quickly, so that in a day or two they are well.

ENLARGEMENT OF THE EPIDIDYMUS AND SPERMATIC CORD.—This disease often follows inflammation of the testicle, and may impede the transmission of the semen to its proper receptacles.

SYPHILIS.—HISTORY OF THE SYMPTOMS OF VENEREAL DISEASES.—I have already stated that venereal diseases are divided into two orders; to the first, I have sufficiently directed the attention of my readers. There remains, however, the second division, which is vastly of more importance, if considered in relation to the general and constitutional results, which are the direct consequence of the primary affection.

PLATE 15.

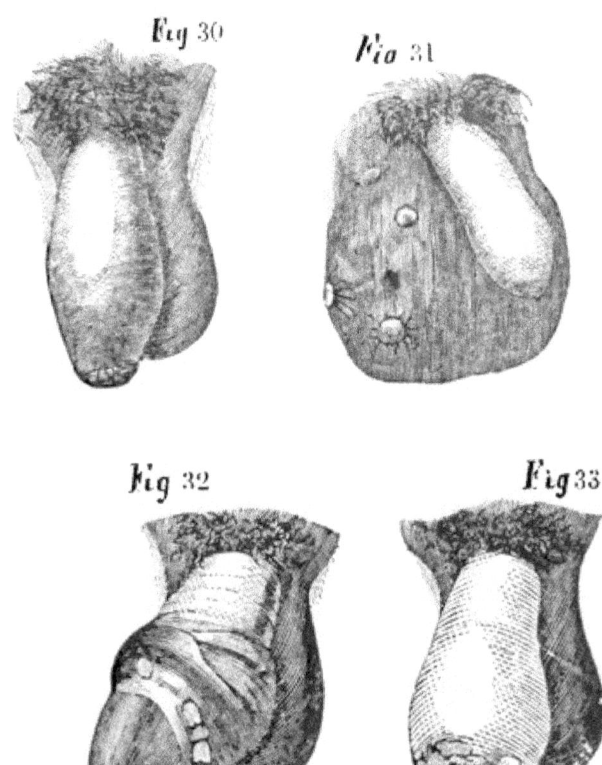

Syphilis is a virulent affection, the essential character of which is its dependence upon a *special* cause, or a distinct *virus*. The *first stage* includes *primary symptoms*, as Chancre, (a pimple or ulcer,) the *specific* cause, from the *special virus* or poisonous matter which has been deposited. The *second stage* embraces *constitutional symptoms*, which follow as a consequence of absorption of the virus, and which is hereditary, and in my opinion capable of transmission by inoculation. Example, various affections of the skin, and mucous membranes, *Enlargement* of the *Glands*, Scrofula, etc. The *third stage* comprehends *Tertiary symptoms*, which can be transmitted, and is hereditary. This is the stage that principally affects the bones.

The destructive effects of the Venereal Disease is becoming so generally known, that it is unnecessary to describe them minutely, though there are many persons even yet, who do not estimate the full extent of their direful consequences on health, reproduction, and longevity. A close and extensive observation in different institutions, established for the cure of venereal diseases, as well as a very extensive private practice for many years, has enabled me to bring the treatment of the various forms of these formidable maladies, almost to perfection. My present purpose is not to give a minute account of all the ravages of these horribly disgusting and malignant complaints, I shall confine myself, therefore, to a few general remarks on their primary and constitutional effects on the human body, in the different conditions of life.

Primary Symptoms.—The first appearance of Syphilis, is a small vesicle on the glans, prepuce, or other part of the penis and testicles of the male, or on the labia, vagina, or uterus, of the female. This is called a Chancre. It arises from the application of the syphilitic virus, on a delicate or abrased surface, from which it is speedily absorbed, in the same manner as the virus of a rabid animal, the virus of small-pox, or of vaccination—only not in so rapid a manner—is conveyed into the body. The whole system becomes sooner or later infected, and a vast number of diseases are developed. Amongst these are Buboes or Venereal Swellings of the glands of the groin, ulceration of the throat, a vast number of cutaneous eruptions, which at first are generally of a copper color, though they may assume the natural appearances of ordinary skin diseases. These symptoms are accompanied or succeeded by pains of the shin and other long bones, as the arms, and even the bones of the head, which are greatly aggravated at night, (I now have two patients that were so situated,) so as to prevent sleep, destroy the appetite and general health, and are often followed by inflammation and swelling of some portions of the periosteum, most commonly on the tibia or shin, instep, back, or palm of the hand. Bones thus affected are termed Nodes.

In numerous cases there is partial or total destruction by ulceration of the virile members, and of the female genitals, of the palate, cartilages of the nose, warts, vegetations or excrescences on the glans penis, or labia pudendi, various abscesses, pustules and fissures in different parts of the body; there are nervous, neuralgic and rheumatic pains, falling off of the hair, phthisis, and very frequently death closes the scene. Vision is often destroyed by the form of ophthalmia, (iritis,) there are severe

PLATE 16.

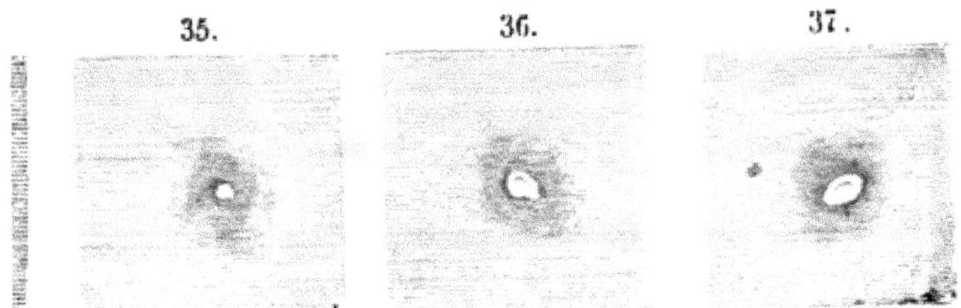

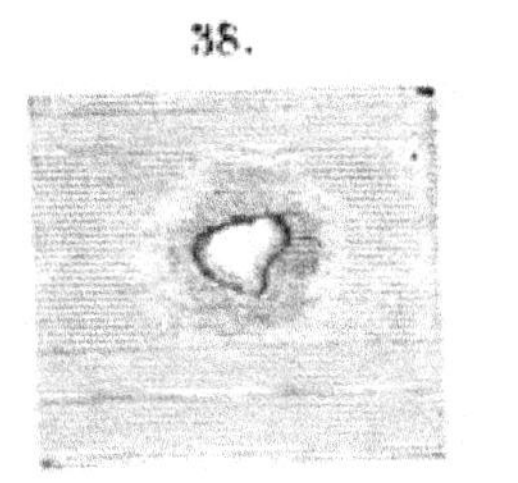

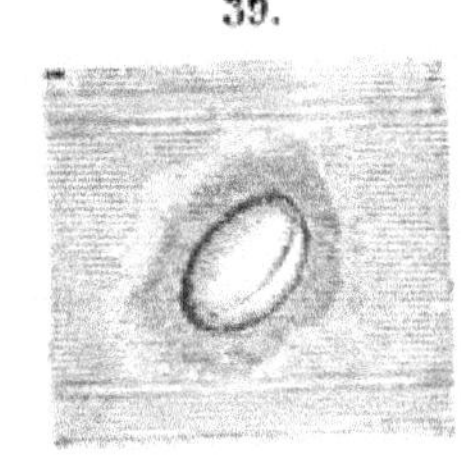

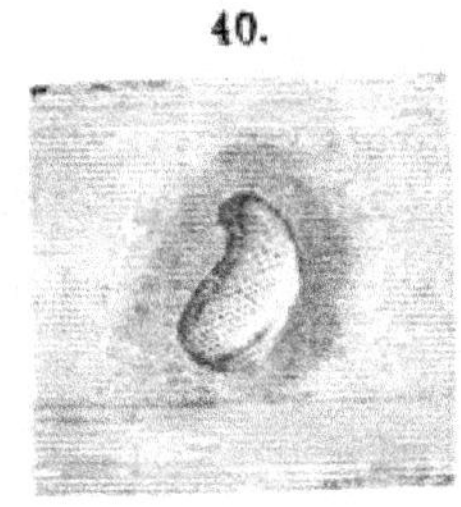

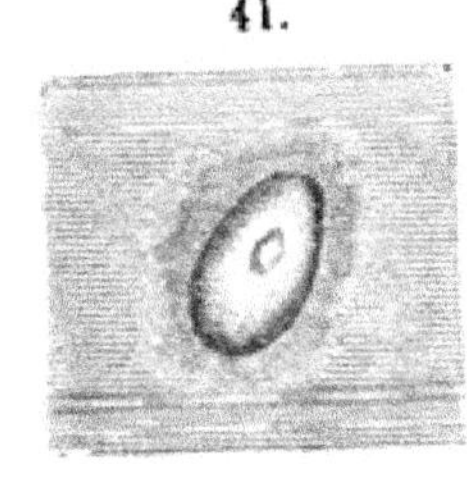

pains in the bones, enlargement termed exostosis, and sometimes caries or mortification, and at other times, brittleness of the bones, which cause them to fracture on the slightest occasion.

The ravages of Syphilis are often hideous, and when the destruction has become extensive, is sometimes incurable. In other cases, supposed to be cured, the disease remains latent in the constitution for many years, is transmitted to the offspring, or destroys the fœtus in the womb. Sometimes it causes sterility. We often observe this disease inducing infecundity and death. In some cases, there is ulceration of the parts, between the bladder and vagina, and the latter and rectum, so that the urine and fœces are evacuated through the vagina, forming a most loathsome and painful disease.

When the venereal contamination of either parent is very considerable, though not apparent, the infant will be born dead, between the seventh and eighth month, in a state of putrefaction. It often happens that women have six or eight infants in rapid succession, which are born dead and decomposed, in consequence of partially cured syphilis in the father. A man who has no external sign of syphilis, and who has been declared cured by his physician, and advised to marry, may contaminate his wife and offspring in various degrees, so that his infant may be born feeble, or covered around the genitals or mouth with red or dark copper-colored eruption. This may appear soon after birth. In primary syphilis of the mother, the infant will be liable to come in contact with the venereal sore, which would cause a chancre on the lip, eye, etc. A sore on the lip of the infant would infect the nipple of a healthy woman, who would then infect every one touching it. Gonorrhœal ophthalmia, will be contracted by the infant during its nativity—if the mother is infected—and will lose its sight in a few days unless proper and prompt treatment is adopted.

A female can, and should be cured of gonorrhœa when pregnant, so that the infant will not be affected, (I now have a case of the kind,) but the fœtus cannot be purified of hereditary syphilis. The mother, however, should be treated without delay.

A man may be relieved of a gonorrhœa, and be affected in no other way, excepting a slight gleet, or a thin watery discharge from the urethra, for years, but if he gets married or cohabits with a healthy woman, he will infect her. The frequency of venereal complaints is much greater than the public imagine. I very often cure boys and girls that have scarcely arrived at puberty even. Young and diffident persons often conceal their situation, until their disease becomes alarming, and then the sufferer will probably apply for advice to advertising empirics, or use quack medicines, which, in nine cases out of ten, allow the disease to poison or destroy the constitution. The proper advice should be sought, and medicines taken on the first appearance of the disease, if the afflicted would save themselves from suffering.

Transmission of the Venereal Virus.—The venereal virus may be transmitted in a few hours, and not in many days. When it is not absorbed at the time of coition, immediate washing of the parts will often prevent infection. I say, if it is not absorbed at the time of coition. What I mean by

PLATE 17.

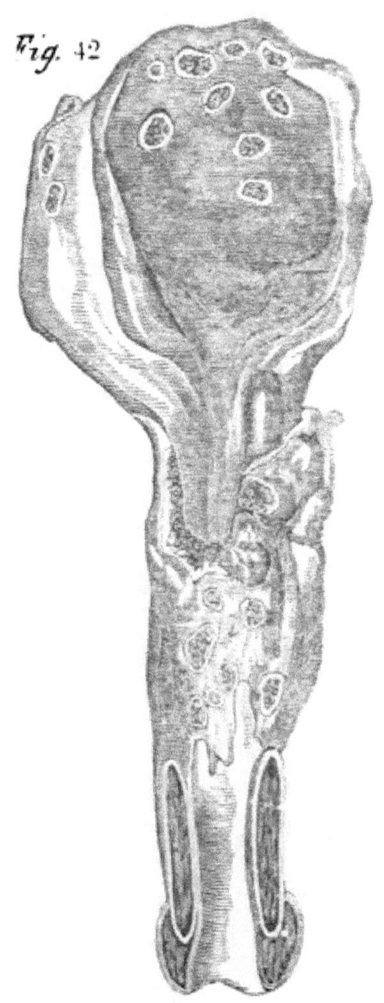
Fig. 42

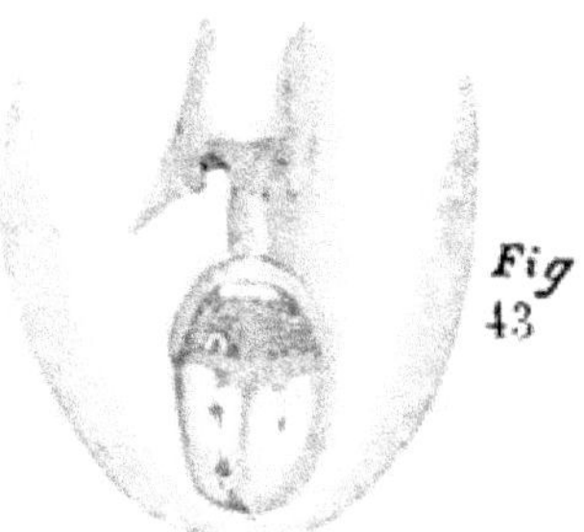
Fig 43

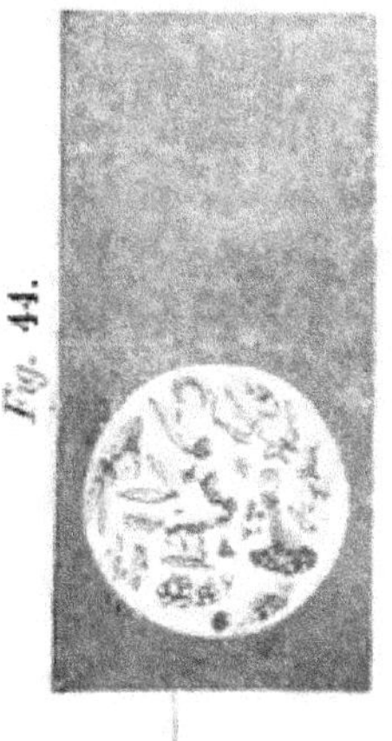
Fig. 44.

this, is, that if the penis is washed before it becomes flaccid, (erection goes down,) infection will often be prevented; but, if the poisonous matter has entered the little follicles or pores of the skin, which are opened in full erection, and close—shutting it in as the penis becomes flaccid—it cannot but be plain to every one, that it is utterly impossible then to prevent the development of the disease. The quack washes, therefore, to be used as an injection for the urethra, or external lotion, are mere humbugs, and worse than useless. Wearing a covering for the penis, is the only thing that will prevent inoculation.

The conveyance of any venereal matter to a mucous surface, such as the lip, eye, nostril, anus, nipple, or to any part where the skin is tender or broken, will communicate the disease. Excoriations of the glans penis, prepuce and labia, are easily distinguished from chancres, and are mere local affections which cannot contaminate the system. Venereal diseases can be contracted from water-closets, privies, from paper or metalic money, washing in the same wash-dish, wiping on the same towel, or using the same lather-brush and razor. I have cured patients who contracted the disease as above stated. As soon as a pimple or little blister has formed, after an impure connexion, on any part of the genitals, it should be properly treated at once.

SECONDARY, OR CONSTITUTIONAL SYPHILIS.

When venereal ulcers or eruptions appear, after a primary sore, on any part of the body, as the face, throat, chest, back, thighs, etc., the constitution is affected, and a judicious treatment must be employed at once, and continued for months after all symptoms have entirely disappeared. The actual length of time necessary to continue the use of medicines varies in different persons, and no one who is not experienced in this class of diseases can safely give advice regarding it.

It is important to distinguish pseudo-syphilis from the real disease, which is often a very difficult matter.

Persons will have secondary syphilis and not have buboes. The symptoms of such a constitutional affection may show themselves in eight days, and may not in years, as I have stated. The term secondary is used to designate the morbid phenomena which appear on the skin, mucous membrane, eye, testicles, etc. When the skin is at first affected, it may appear like measles. The spots, however, will soon lose their rosy color, and assume a coppery hue, or they may disappear altogether. It may first appear in little pimples, or have a scaly appearance, or like chicken-pox, or the pustular form, which is generally very chronic in its course. This form affects the health more than any other. There is another variety which appears on the scalp. A crust is formed around the roots of the hair, which, as often as rubbed off, is reproduced by a thick viscid secretion, matting the hair together; it is usually confined to a few spots, but the whole hair becomes affected, loses its lustre, gets dry, falls off, and the patient may become bald. The glands in the neck may often be enlarged; it often accompanies the other forms of secondary syphilis.

The tubercular form may be little hard tumors, or become ulcers. It may often be seen on the face, nose, or at the angles of the mouth, around the anus, labia, groin, scrotum, lining of the prepuce, umbilicus, between the toes, and in the arm-pits. When ulcerated, it secretes an acrid matter, which causes irritation, and produces an offensive odor; it may remain stationary, or rapidly extend.

SYPHILITIC AFFECTIONS OF THE MUCOUS MEMBRANE, MOUTH AND THROAT.

Every portion of mucous membrane which the eye can observe is like the skin, subject to become the seat of secondary symptoms, as the lips, inside of the cheeks, tongue, fauces, and throat; not only the margin of the anus, but the inside of the intestine itself, also the lining of the prepuce. The vulva, vagina, and neck of the womb are also affected. The throat, from its functions, is frequently exposed to changes of temperature, and feels the effects of all excesses. A primary, as well as secondary affection of the mouth and throat, may exist. The symptoms of these affections, in the commencement, are only slight irritation or swelling, but if inflammation follows, the usual symptoms of sore-throat will occur, and the general symptoms may be severe.

TERTIARY SYPHILITIC AFFECTIONS.

This is a more advanced form of syphilis than any we have heretofore mentioned. Most frequently (in shattered constitutions, or in persons reduced by the combined effects of dissipation and bad treatment) some pain is felt in the throat or tongue; there is a thickness of the speech, which at first excites but little attention, but it will soon expose a tawny-colored ulcer, which may expose the bone, or the palate-bone may be destroyed, and a communication exist between the mouth and nose. The whole back upper part of the mouth will be more or less affected. Not unfrequently pustular eruptions, forming scabs, appear on the extremities, and the general emaciation continues; the countenance has now a cadaverous appearance, and the pulse bespeaks the general feebleness of the patient, who, if not relieved by proper treatment, sinks under the combined effects of colliquative sweats, diarrhœa, great suppuration, and want of sleep, from severe pain in the bones and joints, and loss of appetite. Such is a concise sketch of the most frequent form of tertiary syphilitic sore-throat, with its accompanying symptoms, not to be mistaken when once witnessed. Sometimes little hard tumors, varying in size from that of a pea to a hazel-nut, which, if not properly treated, will soon proceed so that the speech will be very much interfered with.

SYPHILITIC AFFECTIONS OF THE EYE.

The eye, like the skin and throat, may become affected by syphilis in each of its different forms—primary, secondary and tertiary. The importance of the organ, and the rapidity with which the disease can destroy the tissues composing it, deserve the particular attention of the reader.

This affection is usually ushered in by considerable constitutional disturbance, headache, inability to sleep from constant pain over the brow, which is aggravated in damp weather, and in the evening. The eye cannot bear a strong light, and there is usually more or less redness of the ball or lids. It

PLATE 18.

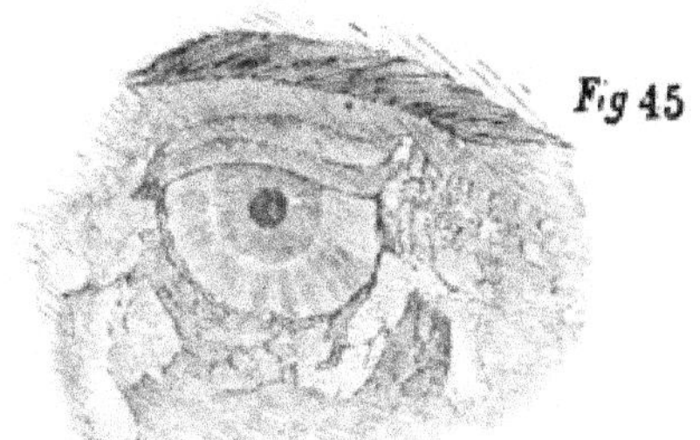
Fig 45

Fig 46
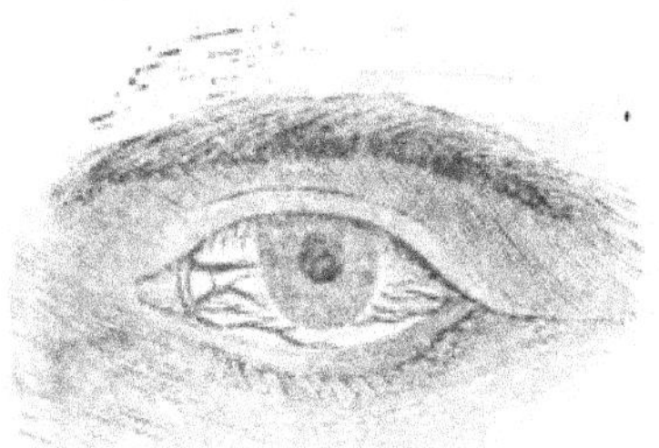

Fig. 47
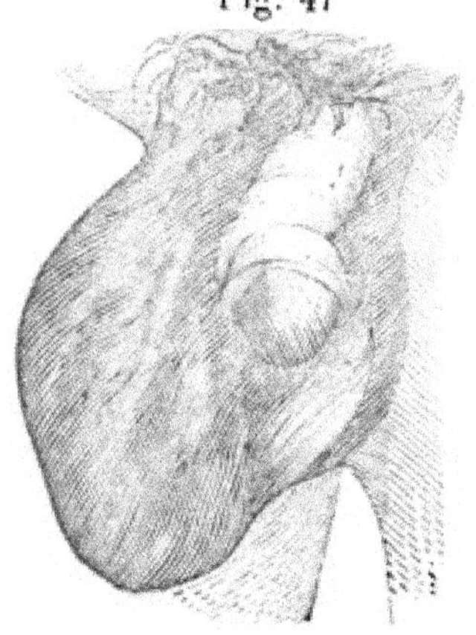

may appear on the eyelids in the form of ulcerations.

In a case of direct inoculation of the eye by the primary syphilitic virus, which often happens by carelessness, in not keeping the hands washed clean, or wiping them on a cloth, which should be destroyed after such use. The affection will present nearly the same appearance and condition as a chancre on the genitals, only the discharge from the sore will rapidly spread over the whole eye, and if the patient is not careful while lying down or asleep, it will come in contact with the other eye, and both will be destroyed if the disease is not immediately checked.

SYPHILITIC AFFECTIONS OF THE TESTICLES.

Some months after the occurrence of the primary symptoms, the patient will complain of vague pains in one or both testicles, particularly felt toward night, and which shoot upward toward the loins. In some instances no pain is felt, and the patient is surprised at finding one or both testicles gradually enlarging, with a very considerable inconvenience from their weight. The functions of the organs will become impaired if the disease is not speedily arrested, and the correct treatment followed up for months, so that the disease will be thoroughly eradicated from the system.

TERTIARY SYPHILIS.

This constitutional affection is included under the name of *Nodes*, inflammation of the periosteum, exostosis, càries, and tubercles. Its destructions are principally confined to the bones. When the cellular tissue is affected, small tumors will first be observed, either isolated or occurring on various parts of the body. They may not be noticed much for some months. They may be hard and unattached to the adjacent structures. It may make its way to the surface, ulcerate, and then heal, but no sooner has one disappeared than another shows itself.

TERTIARY SYMPTOMS AS THEY OCCUR IN THE OSSEOUS SYSTEM.

The first symptoms of the affection of the bones, consists in pain at first vague, or like rheumatism; they will generally become fixed to particular bones sooner or later, and become more severe at night. There are three varieties of *Periostitis*.

Ostitis, shows the same preference for particular regions. After having existed a long time without giving other indications of its presence than pain, the swelling it ultimately gives rise to betrays itself externally. In other cases it may give rise to *parenchymatous*, *exostosis*, or *hyperostose*.

In the second part of this work will be found a great number of illustrative cases of all the diseases which have been mentioned, in the shape of certificates of cures, and I shall pass on to that part.

Enough, I think, has been shown to convince the most unthinking individual who may peruse these pages, of the evils and miseries induced by venereal abuses, excesses, and diseases.

PLATE 19.

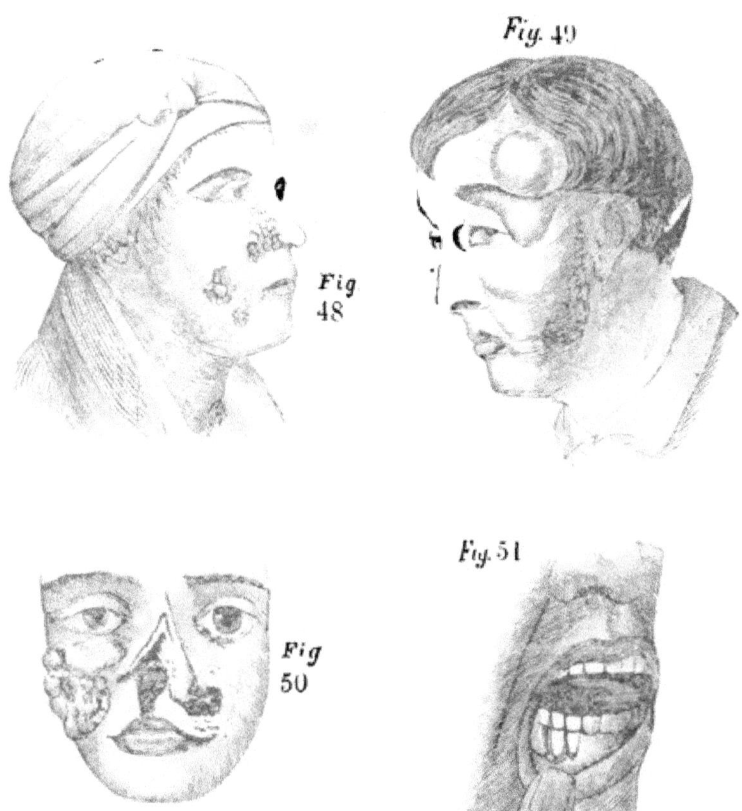

Special Notice of Injuries Received by Patients from Inexperienced Physicians, Quacks, and their Remedies.

As we have but this additional page to give the statements of a few more patients, of the injuries they received from treatment they had submitted to previous to applying to us, we merely give their initials, and the most dangerous symptoms they were laboring under at the time of such application. All of them can be seen in full, at our offices, but of course without their names.

November 15, 1858.—A., Georgia. I have tried this physician, and that, using copaiba, nitrate of silver, and other caustics for seventeen months, till I almost despaired of being cured. I have nocturnal emissions, continual burning in, and discharge from the urethra; my semen is as thin as water, and can only partially cohabit with a woman. J. L. O.

October 26, 1858.—I have a continual desire to pass urine, and cannot retain it; am totally unable to cohabit with a woman; loose semen nights, with urine, and at stool; I paid about $150 for Tricsemar, but received no benefit. C. F., N. Y., April 3. 1858.

W. N. B., N. Y., was treated for nine months by one of New York's most eminent professors without benefit. but we cured him in about a month.

January 19, 1858.—R. T. B., Va., applied with nocturnal emissions weekly; total paralysis of the penis; muddy and shreddy urine; pain and constant desire to pass it, day and night; yellowish discharge from the urethra, after using acids or stimulents, or with a cold; scrotum relaxed, and testes swollen; sleeplessness, giddiness, acidity of and wind in stomach, and bowels; gloomy. Had been confined to the house a number of months, though he had been under the treatment of a number of the best physicians, more or less, for years. We cured him in a few months.

November 30, 1857.—W. R., Iowa, came with nocturnal emissions every week; weak back; a tea-spoonful of a sandy looking sediment in the urine that stood over night; constant desire to pass urine night and day; partial erections; iritis; had been cauterized a number of times; cured by us in four months.

PART II.

PARTICULAR CAUTION.

Avoid all quack or advertised remedies for any disease, especially venereal, and all of the instruments advertised to cure Seminal Emissions, as they are injurious quack humbugs.

Within the past year or two, there have arisen a number of mushroom doctors, pretending to be from Hospitals in the different cities of Europe, or students of eminent men, such as Ricord, Acton, etc., etc., using their medicines,- medicated bougies, rectum suppositories, etc., etc., whose injured and uncured patients we are curing every week. Another caution, to save yourselves from imposition in our own building, be certain you see our No., 647 Broadway, and the silver plates on each side of the door, stating the "Physician's Consulting Rooms are on the second floor," where, upon ascending the stairs, as directed, you will see our names upon silver door plates, on the doors of the offices.

PREFACE.

This work has been hastily prepared, in fact, was not commenced until within the last few months, except the cases taken from my note-book, and has been prosecuted amidst the most pressing professional engagements. Notwithstanding these disadvantages, I flatter myself that it will answer much of its intended purposes. These purposes are, mainly—

To place before the reflecting public as unexceptionable a work as may be, compatible with the unpleasant topic treated:

To supply a desideratum in popular medical literature, great care has been taken to simplify the subject in plain language, divested of technicalities and exceptionable phraseology:

Avoiding everything calculated to excite the passions or administer to an impure appetite for vulgar books, which, however, seems not to have been the aim of most writers on kindred subjects:

To, finally, introduce to the family and social circles in a modest garb, a timely Adviser against the deplorable consequences of Onanism—that most fatal and pernicious habit of youth—that dreadful scourge of humanity—that untimely Destroyer—

> " He, the young and strong, who cherish'd
> Noble longings for the strife,
> By the road-side fell and perish'd,
> Weary with the march of life!"

By thus warning, I do not purpose to cast a gloom over the minds of sensitive sufferers, but to promise hope, relief and perfect restoration, even in the most unhappy state of extreme debility; from my vast experience and truly wonderful success in curing the most unpromising subjects of Spermatorhœa and other diseases of a private nature—

> "The miserable have no other medicine,
> But only hope."

But it is always advisable to seek relief early, thereby avoiding much unnecessary suffering.

The deplorable condition of the nervous system induced by the depressing, degrading, demoralizing and disorganizing habit of Self-Pollution, in so many cases. causes that unhappy irritability of temper, melancholy dejection, moroseness, and finally that intellectual imbecility which terminates in an awful wreck of the mental and physical organism. It is not the object of the author to paint the effects of Onanism in the deepest dyes, to horrify young imaginations; but to warn—to-day—now—for to-morrow the shafts of disorganization may have been sunk too deeply into the system, ever to be removed by the Healing Art.

New York, 1852. THE AUTHOR.

INTRODUCTION.

THROUGHOUT the present work, to prevent too frequent repetition, I shall use Spermatorhœa, to designate the disease consequent upon the habit of self-abuse; in short, any undue loss of semen, from whatever cause.

A sufficient number of cases will be inserted to illustrate the various effects of Onanism upon the different temperaments, or the peculiarities of age, sex, or constitutional predisposition to diseased action.

The afflicted cannot expect to obtain, from any work on medical subjects, a sufficient knowledge of disease, in its different phases, complications and results, to be able to treat themselves.

From the intimate association of one organ and its functions with others, and from the harmony existing throughout the whole frame, a part of the animal economy, however small it may be, cannot scarcely be disordered without causing deranged action in other parts; no modification can occur without involving in its changes other organs, and disordering their functions also.

Even medical men, generally, are wholly incompetent to advise and treat successfully the affection under consideration.

It may not be egotistical for me to remark here, that am fully aware of the importance of the subject, and that, from my long and special attention exclusively to the treatment of thousands of patients afflicted with

Spermatorhœa and other sexual complaints, and that, too, with the most flattering success; that it is probably as well that this task was left to me by the profession.

The public have not the least conception of the vast amount of suffering entailed upon humanity by the practice of self-abuse, and the too free indulgence of the baser passions. My readers, who will follow me to the end of this modest little volume, will be amazed at the sad effects of what they may have considered an innocent practice.

Nor should I, in the outset, hesitate to repudiate the pernicious conduct of quacks and low-minded men, who have written books purporting to be guides to the young, but, instead, have been inflammatory appeals to their passions. And not a few of these books plainly inculcate immorality, for the sake of administering to the impure desire for bawdy books, too common among youth of both sexes. The great aim of this work, therefore, is to rescue the young and ignorant from the frightful abyss awaiting them, if deprived of a warning voice. The author also deems it his duty to denounce the introduction of plates representing the genital organs of both sexes, as improper for a popular treatise.

If ladies and gentlemen wish to study Anatomy and Physiology, we refer them to popular school-books upon those subjects, as a healthy and proper study for the most fastidious; and not to resort to a book of bawdy, incorrect and intrusive plates, in senseless and ridiculous books, issued as an advertising medium for Quack nostrums, and, still worse, to ignorant pretenders, who set themselves up for Doctors.

We here take a very important step, in warning our afflicted readers against applying to villainous impostors who infest large cities, and who not only swindle the innocent patient of his money, but leaves the system a wreck to poisonous nostrums.

Another reason that induced me to send out into the world this volume, (I am sorry to say it), is the almost total ignorance of the members of our most benevolent Profession, many of them not knowing such a disease as Spermatorhœa. And I do hope, for the sake of humanity, the medical profession generally will not rest a single day, until that knowledge is obtained; as three-fourths of all the chronic diseases have their origin in Onanism.

It is a deplorable fact, that young ladies, as well as youth of the opposite sex, are addicted to the habit of self-abuse to an alarming extent. This habit frequently has its origin in irritation of the genitals and lower bowels of children, from costiveness, or from pin-worms, (Ascarides), which are excessively annoying to some children. The frequent inclination to rub or scratch the adjacent parts lead to the practice of Self-Pollution. Many instances of this character have come under my own observation.

I have many times been consulted by mothers, whose daughters have suffered from these abuses, consequent upon irritation of the Pudendum, at the age of puberty.

"Mine honor's such a ring:
My chastity 's the jewel of our house,
Bequeathed down from many ancestors;
Which were the greatest obloquy in the world,
In me to lose."

M. LARMONT

A FEW WORDS TO INVALIDS.

However hopeless you may think your case, whether from extreme debility and prostration of your entire system, or from the unsuccessful treatment of the score or more of Physicians you unfortunately may have been under, or from the false delicacy that many of my patients tell me kept them from an earlier application to me, or the no less fatal idea of your overcoming the disease by the advancement of age, and a strictly moral mind, &c., &c.; for they one and all only make your case worse, and if it is an early one, from the cause being entirely local, it affects the entire animal economy, and then requires general as well as local treatment, which I am happy to be able to say, the science of Medicine and great improvement in the Healing Art, warrants me in saying, that the most severe and obstinate cases, of however long duration, yields to my mode of treatment, as the numerous voluntarily written Certificates contained in the following pages, from persons of all ages and in the highest stations of life, which portray the difficult stages of the disease, as developed in the different constitutions, will fully attest. I give the initials only in most of the cases from delicacy, but the full names by their cordial permission, can be seen at all times in my office, as well as letters from Members of the Medical Profession in this and other cities.

The letters, or names of patients, are known only to myself, as I have no Assistant or Students; notwithstading this, I always destroy them at the termination of each case. Therefore, the unfortunate can disclose their bleeding hearts to one secret bosom whose feeling throb will return that great boon, Health, for the so generously reposed confidence in him.

THE AUTHOR.

PLATE 20.

Fig 52

Fig. 53.

CHAPTER I.

THE IMPORTANCE OF PERFECTLY UNDERSTANDING THE SUBJECT OF ONANISM—ITS EXISTENCE FROM THE EARLIEST PERIOD, AND YET THE IGNORANCE OF THE MEDICAL PROFESSION—IMPOTENCE AND STERILITY, A CERTAIN RESULT IF NOT CURED—THE EFFECTS, OF IMPOTENCY AND STERILITY ON THE HUMAN RACE.

Why is this an important subject? It is because the commencement of this habit is the laying the foundation of almost every organic disease, the vital system is subject to. Any person on a moment's reflection, will be convinced of the intimate relation existing between the Mind and the Genital Organs, and their intimate connection with the whole nervous system. What renders it of more importance still, is, there is scarcely anything known about it, by any one but the afflicted, as in the early studies of medical men, they are taught nothing of it, and in their later years, their time is so fully occupied with a general practice, they have not the opportunity requisite for an understanding of such a speciality, and yet, as regards health and happiness, there is none that possess one-twentieth part of the importance. To convince you that I do not speak without authority, I here insert an extract from Dr. Curling's remarks, when speaking of these organs:—

"Their functions being so involved in those of other parts, are influenced by such peculiar causes, and are so dependent on and modified by particular events and circumstances, that the investigation of them when disordered, necessarily becomes of a complex and difficult character. The product too, of these glands, is one, the qualities of which it is almost impossible to appreciate, and which during life is never afforded in a pure and unmixed state; and further, taking into account the repugnance felt to such inquiries, it is scarcely surprising that the subject has been but imperfectly investigated, and rarely treated of by the pathologist and practitioner Indeed, the little information we possess respecting it, is chiefly to be found under the head of Impotency, in works on medical jurisprudence, in which it is cursorily considered, principally in relation to points of Medico-legal interest, and scarcely at all in reference to practice."

This certainly is the truth, and it no doubt will surprise many persons to learn, that the generality of physicians know little or nothing of these matters. I often have patients who have applied to the most eminent physicians and surgeons, of the largest cities in the world, and paid them consultation-fees, to be told there was nothing the matter with them. It is only a few weeks even, since a patient came to me, and who, among other great men he had consulted, was Sir Benj. Brodie, of London, and whose directions were not to use any of their dirty beer, and he would be well enough; that his disease was only imaginary, and if he would only think so, he would have no farther trouble. You can imagine the gentleman's feelings, as he was a strictly temperate man. From that time he began to despair, for he had lost almost all the power of the organs. What was more sad still, he had not the ambition nor desire, until he put himself under my charge. I am happy to say, however, that in a few weeks he was entirely in the full possession of that vigor, both of body and mind, which nature had formerly bestowed upon him.

The statements of their individual cases by the patients themselves, and their certificates for their cure by me, will be found in the subsequent chapters, arranged under their proper heads, according to their advanced or primary stage which they had reached, previous to my treatment. It will be sufficient, I opine, to not only illustrate the want of knowledge of the medical profession as to the proper treatment for such cases, but whether my own, is not superior to that of any other. From the want of knowledge as heretofore mentioned, there is very little, if any, positive knowledge of these complaints being known as distinct diseases, in that catalogue which enumerated so many calamities with which our ancestors were afflicted. But we can infer to a certainty, I think, that it did exist for centuries, previous to its having become known that it was a distinct disease, when we are made acquainted with a fraction only of the immense amount of suffering, at the present time, among the modest, diffident, pious youth, who, unconscious and ignorant of doing harm, when they first learned the practice at the seminary, academy, college, or from some of their acquaintances while at home. Josephus mentions this as a disease, when relating the ancient purification laws of the Jews. He says, "He that sheds his seed in his sleep, shall be privileged with those who have wives." In relation to another disease, he gives us another of their laws: "Those who had a Gonorrhœa, were prohibited from coming within the bounds of the city."

This not only bears witness to there having been such a disease known at that time, but it showed their wisdom was far superior to many of the pretended medical savans of the present day; for they adopted a regulation, which they supposed would be a remedy for all such cases.

Some have referred the origin of this abuse to the idolatrous worship of the northern Venus, named Frago, in oblation to whom her votaries were accustomed to shed their seed. The opinion of the All-wise, upon the enormity of this offence against reason and nature, will be found where he speaks of this dereliction of *Onan*, in Genesis, chapter thirty-eight, ninth and tenth verses:—

"And Onan knew that the seed should not be his, and it came to pass, when he went in unto his brother's wife, *that he spilled it on the ground*, lest that he should give seed to his brother. And the thing which he did, displeased the Lord, wherefore he slew him also."

It is only a few years since it was thought degrading for any one even to allude to masturbation, and especially to speak openly or write about it. But as progression is the spirit of the age, people have become more rational, and are beginning to know the practice is one of the most injurious, possible; and that to be able to remove the disease, it must of course be understood. As I have before stated, however, the books heretofore published on this subject, were entirely inadmissible to families. This work obviates that objection, and yet has entered sufficiently into the physiological and medical details to give a clear view of the evil From the extracts which I give below, from the ancient medical writers, the reader will be able to see how this subject was regarded by eminent minds, in former ages. A number of these extracts are from M. Tissot and others. They are from the writings of the Fathers of Physic, many of whom lived centuries before Christ. Allowance must be made for the deficiency in scientific accuracy, for the whole truth could not be told as it was not known.

Hippocrates, the oldest and most correct observer, has already described the diseases produced by abusing the pleasures of venery, under the term of dorsal consumption. "This disease," says he, "arises from the dorsal portion of the spinal marrow. It principally attacks young married people, or the licentious. They have no fever, and although they eat well, they grow thin and waste away. They have a sensation like ants crawling from the head down along the spine. Whenever they go to stool, or evacuate their urine, a considerable quantity of very thin seminal fluid escapes from the urethra. They lose the power of procreation, yet often dream of venereal pleasures. They become very weak, and walking produces shortness of breath; they have pains in the head, and ringing in the ears, and finally, an acute fever (Libiria,) supervenes, and they die." Some physicians have ascribed to the same cause, a disease which he has described in another place, and have termed it the second dorsal consumption of Hippocrates, and which has some relation to the first. But the preservation of the strength which he mentions particularly, seems to us a conclusive proof that this disease does not depend on the same cause, but seems rather to be a rheumatic affection. "These pleasures," says Celsus, in his excellent work on the preservation of health, "are always injurious to weak persons, and their abuses prostrate the strength."

We can find nothing more frightful than the description by Aretaus, of the diseases produced by a too abundant evacuation of semen: "Young persons assume the air and the diseases of the aged; they become pale, stupid, effeminate, weak, idle, and even void of understanding; their bodies bend forward, their legs are weak, they have a disgust for everything, become fit for nothing, and many are affected with paralysis." In another place he mentions the abuse of these pleasures as among the six causes which produce paralysis. Galen has seen diseases of the brain and nerves from the same cause, and the powers of the body impaired; and he also relates that a man who was convalescent from a violent attack of disease, died the same night after coition with his wife. Pliny, the naturalist, informs us, that Cornelius Gallus, the old prætor, and Titus Etherious, a Roman knight, died in the act of copulation. Aetius says, "the stomach is deranged, all the body wastes, becomes pale, dry, and the eyes sunken." These remarks, of the most respectable ancient writers, are confirmed by the moderns.

Sanctorious, who has examined with the utmost care, all the causes which act on our bodies, has observed, that "this weakens the stomach, destroys digestion, prevents insensible perspiration, the derangement of which produces such evil consequences, disposes to calculous diseases, diminishes the natural warmth, and is usually attended with a loss or derangement of sight."

Lomnius, in his fine commentaries on the passages of Celsus, whom we have just cited, supports the remarks of the author by his own observations: "Frequent emissions of semen relax, weaken, dry, enervate the body, and produce numerous other evils, as apoplexies, lethargies, epilepsies, loss of sight, trembling paralysis, and all kinds of painful affections."

One cannot read without horror, the description left us by Tulpius, the celebrated burgomaster and physician of Amsterdam:—"Not only," says he, "the spinal marrow wastes, but the whole body and mind become languid, and the patient perishes in misery. Samuel Vespertius was attacked, first with a humor upon the back of his neck and head; it then passed to the spine, to the loins, to the lower and lateral region of the abdomen, and to the hips. This unhappy man was affected with so much pain that he was entirely disfigured, and was emaciated so gradually by a slow fever, that he more than once asked to be relieved from his misery by death."

"Nothing," says a celebrated physician of Louvaine, "weakens the system so much." Blancard has known simple gonorrheas, dropsies and consumptions to depend on this cause; and Muys has seen a man of good age attacked with spontaneous gangrene of the foot, which he attributed to the same kind of excesses.

In the Memoires des Curieux de la Nature is mentioned a case of blindness, which deserves to be given at length. "We are ignorant," says the author, "what sympathy the testicles have with the body, but particularly with the eyes."

Salmuth has known a sensible hypochondriac to become a fool, and in another man the brain to become so collapsed that it was heard to rattle in the cranium, both from excesses in venery. I have known myself a man, fifty-nine years of age, who, three weeks after marrying a young wife, became blind, and in four months died.

The too great loss of the animal spirits weakens the stomach, and destroys the appetite; and nutrition not taking place, the action of the heart becomes more feeble, all parts languish, and the patient becomes epileptic. It is true, we are ignorant whether the animal spirits and the seminal fluid are the same; but observation shows, as we shall see hereafter, that these two fluids are very analogous, and that loss of the one or the other produces the same complaints.

Hoffman has seen the most frightful symptoms ensue from the loss of semen. "After long nocturnal pollutions," says he, "the patient not only loses strength, becomes emaciated and pale, but the memory is impaired, a continual sensation of coldness affects all the extremities, the sight becomes dim, the voice harsh, and the whole body gradually wasted; the sleep, disturbed by unpleasant dreams, does not refresh, and pains are felt like those produced by bruises."

In a consultation for a young man, who, among other diseases produced by masturbation, was affected with weakness in the eyes, he says, "I have seen several instances of young men, who, at mature age, when the body possesses all its strength, were attacked, not only with severe pain and redness of the eyes, but the sight became so feeble, that they could neither read nor write." He adds, "I have even seen two cases of gutta sereno, from the same cause."

The history of the disorder which gave rise to this consultation will be read with interest: "A young man commenced masturbation, when fifteen years old, and having indulged in it till he was twenty-three, experienced so great feebleness in his head and eyes, that during the emission of semen there was severe pain in the latter. When he attempted to read anything, he had a feeling similar to that of drunkenness; the pupil was extraordinarily dilated; the eyes were exceedingly painful; the eyelids very heavy, and glued together every night; they were often filled with tears, and a whitish matter collected very abundantly in the two corners, which were very painful. Although he ate with a good appetite, still he was extremely emaciated; and after he had taken food, appeared as if drunk."

The same author has mentioned another case of which he was an eye-witness, and which, we think proper to mention here:—"A young man, eighteen years old, who had had frequent connections with a servant girl, suddenly fainted, and trembled exceedingly in all his extremities; his countenance was red, and his pulse very small. He recovered from this state at the end of an hour, but continued very feeble. The same phenomena occurred very frequently with severe pain, and at the end of eight days there was a contraction and tumor in the right arm, with a pain in the elbow, which was always increased during the paroxysm. The disease increased for some time, but was finally cured by Hoffman."

Boerhave portrays these diseases in that masterly manner and with that precision which characterizes all his descriptions:—"Too great loss of semen produces weakness, debility, immobility, convulsions, emaciation, dryness, pains in the membranes of the brain, impairs the senses, particularly that of sight, gives rise to dorsal consumption, indolence, and to the several diseases connected with them."

The cases narrated by this great man to his auditors in explaining to them this aphorism, which related to the different kinds of evacuations, ought not to be omitted:—"I have seen," says he, "a sick man where the disease commenced by a lassitude and feebleness in the body, particularly in the loins; it was accompanied by twitching of the tendons, periodical spasms and loss of flesh, so as to destroy the whole body; also pains in the membranes of the cerebrum, pains which the patient terms a dry burning, (ardeur seche), which constantly inflames this most noble organ. I have also seen one young man affected with dorsal consumption. His figure was good, and although often cautioned against indulging in these pleasures, he did not regard it, and became so deformed before death, that the layer of flesh which appears above the spinous processes of the lumbar vertebræ, entirely disappeared. The cerebrum in this case seemed to be consumed; in fact, the patient seemed to be stupid, and became so stiff, that we have never seen the body so immovable from any other cause. The eyes are so dull that the sight is nearly lost."

De Senac mentions in the first edition of his Essays, the dangers attending masturbation, and states that "all who indulge in this vice will be affected in the flower of their youth with the infirmities of age." We can see in the following editions why this and other remarks of the same character were suppressed.

Ludwig in describing the diseases resulting from too frequent evacuations, does not forget that of the semen. Young people of both sexes, who indulge in lasciviousness, ruin their health, by wasting strength which was designed to make them vigorous, and finally fall into consumption. De Gottier details the sad accidents arising from this cause; but they are too long to copy. We refer to the work, all those who can read the language in which it is written.

Van Swieten, after quoting the description of Hippocrates mentioned above, adds—"I have seen all these symptoms, and several others, in those unfortunate people who indulged in self-pollutions. I have employed uselessly, for three years, all the resources of medicine, for a young man who was diseased in consequence of this practice, with wandering, frightful, and general pains, with a sensation, sometimes of heat, and sometimes of cold, in every part of the body, but particularly in the loins. Afterwards, these pains having diminished, his thighs and legs were so cold, that although they seemed of the natural temperature when touched, he was constantly warming himself by the fire, even during the warmest days of summer. I noticed particularly all this—a continued rotary motion of the testicles in the scrotum, and the patient felt a similar motion in the loins." This account does not mention whether this unfortunate creature died in three years, or continued to languish some time longer, which would be more dreadful—he could not have recovered.

Kloekof, in a very fine work on the diseases of the mind, which depend on the body, confirms by his observations what we have already mentioned.

If these emissions, which are sure to be the sequence of such a habit, continue, they at first cause a feeling of lassitude, sleepiness, want of ambition, hesitancy of action and speech, occasional and permanent, even dimness of vision, thoughts diffused, memory treacherous or lost, nervousness, such as being easily frightened, sudden starts at trifling noises, want of confidence, a desire not to mingle in society, especially that of females, impotency and sterility, indigestion, costiveness, or diarrhœa, (generally costiveness). Those most annoying of anything, dark spots under and around the eyes, loss of flesh, pain in the chest and back, weakness, (particularly of the legs), coughing, consumption, idiotcy, and insanity. Now, these are the results, as hundreds of my patients are ready to testify.

And yet there are pseudo-philanthropists who refuse to advertise, to let the world know The physician and surgeon, they think, has demeaned himself, by devoting his time to this particular practice. Grave professors are following the same jack-o'-lantern. That is the reason medical students do not have the subject even mentioned to them, for I have some of them my patients, every few weeks; but that is not all—they have not only left the victims to themselves, but they have established that hydra-headed monster, Quackery, and the grave is now indebted to that abandonment for its thousands.

One of the results, previously mentioned, of this vitiated indulgence, is Impotency, and finally sterility. A person with the most circumscribed imagination, can easily foresee the consequent misery, certain to befall them, if they are so imprudent as to enter into a matrimonial alliance before they are cured. The broken heart of their unsuspecting, and, may be, too confiding partner, I will not say too confiding, for I have never met a single patient that would have entered the marriage state, if they had known what their deplorable situation would have caused. They would a thousand times have terminated their existence rather than the consummation of the marriage ceremony.

I am fully convinced, that a majority of those who commit suicide are impelled to do so by that morbidly diseased state of the mind arising from the habit of masturbation, for when it has proceeded far enough to cause impotency and sterility, the mind is so affected, that reason is overpowered and replaced with the horrible idea that every one they meet is aware of their complaint, therefore the desire of ridding the world of their presence. This is not an ideal picture, by any means, as forty-nine out of fifty of the afflicted can testify to its reality.

CHAPTER II.

THE EFFECTS THE MIND HAS ON THE ORGANS OF GENERATION—STRIKING CASES AS WITNESSED BY OTHER SURGEONS, AS WELL AS MYSELF.

Persons of mature age have more or less experienced similar effects. A libertine, or sensually dissolute character, is a person whose mind is not actively employed in business, or close application of the mind in any pursuit, but one who has been fortunate in inheriting wealth or a competency from their relatives or friends. If they have not been favored in that respect, to enable them to live without exertion, further than the planning of some sumptuous dinner, or the stimulation from liquors at champaigne suppers, sporting, &c., their resort is the gaming-table.

The mind, excited with alcoholic drinks, or the highly

and richly seasoned food, has full scope to anticipate sexual pleasures, but when the labor necessary for the success of business is preying on it, it only anticipates that rest which sleep so well affords, for I have known many and many a person so absorbed in some favorite or necessary pursuit, as to have months elapse without any desire for sexual indulgence, but as soon as the mind was at liberty again they discover their inability, and with the proper treatment, a full restoration. When the organs are in a healthy state, desire can be engendered or dissipated by mental impressions; in fact, the growth and development of the organs can be accelerated or retarded in a degree by the same means

There are many persons who do not feel sexual desires until a late period of life; the consequence is, the organs themselves are not only imperfectly developed, but the body and mind are also retarded in the same way. Sometimes the long suppressed feelings will receive a sudden stimulus, from seeing some person of the opposite sex particularly adapted to make this desired impression upon their minds. Every person of experience very well knows that a certain impression must be made on the mind, before the animal feeling is experienced, or the physical development takes place, for there are many of the opposite sex who excite disgust, and under such circumstances the certain feelings for enjoyment would not only not be produced, but if age with the necessary favorable circumstances, had not already caused a full development of the organs, there would be great danger of their ever fulfilling nature's intentions. This at once explains to us the reason of so many of those distressing cases of indifference and dislike to be met with between parties, and be a partial aid in giving the necessary treatment.

There are good reasons for supposing that the sexual instinct is materially dependent upon a particular part of the brain, though we are not exactly certain what part it is, nor whether it is a mere development of it that is needed, or some peculiarity of structure or organization. It is not at all uncommon to find men perfectly organized in every respect with vigorous minds, and with every other faculty in full play, but yet almost wholly destitute of desire for sexual enjoyment. In most of these cases, it is true, the generative organs are small or inactive, yet in some they are of full development, healthy, and active. In such cases we can only account for the singular indifference exhibited, by supposing that the part of the brain which regulates the reproductive instinct has not had sufficient power, or else the senses have not been properly presented. In the same way, we can account for the influence of the brain and nervous system on the generative organs, as previously mentioned; such as exhausting the nervous energy, if we can so speak, in thinking, or in muscular energy. The other functions, as well as the generative, are proportionally weakened. Authors are apt to become impotent, when the mind is very intently engaged on some particular subject. In the lives of several students, we have a further corroboration in proof of my assertion, as many of them have been remarkable for their coldness and incapacity, and particularly when engaged in such absorbing and abstract studies as Mathematics. As another instance—Sir Isaac Newton is said to have never known sexual ardor, though in every respect a perfectly formed man, and it is probable, I may say without a doubt, it was owing to his incessant and all-absorbing studies.

I have the fullest proof in sustaining me in the assertion, that intense mental occupation will not only lessen the sexual ardor, but that it will extinguish it entirely. This is not a simple fact which should be so easily passed over—as it is of the highest medical and moral importance.

The only thing required in these cases, is, an intelligent and honest physician, who will examine and find the real cause, and then in a sympathizing manner explain it to the patient, administer the proper treatment, and the cure can be effected. But if a wrong course of procedure be adopted, then the evils will be confirmed. It is in this, as in every thing else; ignorance and concealment from the proper physician produce evils that only knowledge and mutual confidence can prevent or remove.

Be cautious in choice of your "Doctor, Dear,"
But after that choice, abandon all fear.

CHAP. III.

INJURIES OF THE BRAIN, CAUSING IMPOTENCY—CASES ILLUSTRATIVE OF THE SAME.

Baron Larrey gives the remarkable case of a soldier, which came under his own notice. He was a healthy, robust man, with strong propensities and endowments, who had a portion of the back part of the head cut off by a sabre. He recovered from the wound, but lost the senses of sight and hearing on the right side. Pain was also experienced down the spine, and a peculiar creeping feeling in the organs, which also began to waste, and in fifteen days the power was entirely lost.

M. Lallemand also mentions a case of a French soldier, similarly injured, in the expedition to Algiers, who speedily experienced the same wasting of the organs, loss of power, and sexual desire.

In the American Journal of the Medical Sciences, for February, 1839, Dr. Fisher relates a curious instance of a gentleman injured in a railway car. He was looking out of the window at the moment when the collision occurred, and the shock threw his head against the edge of it, striking the back of it with such force as to stun him; he, however, recovered his senses, and was taken home, but suffered great pain in the back part of the head and top of the neck. His right arm was benumbed a little, and some difficulty was experienced in passing the urine; but in two weeks he was able to walk out with no other inconvenience than a slight dimness of sight. About the fifth week he discovered that he was impotent, and had lost all sexual desire. The means used to restore his genital powers were only partially successful, nor was his memory so perfect as before, but all the other difficulties disappeared under the proper treatment.

In the Lancet, for August, 1841, is an account of a medical student who received a blow on the face, in a quarrel, which knocked him down, so that he fell on the back of his head. He was totally unconscious for eight or ten hours, but gradually recovered, and on the following day even resumed his studies, which he continued unremittingly for the next six weeks. He, however, became exceedingly irritable, with a feeling of general uneasiness, and after the first week, he observed the genital organs begin to waste and the desire to grow less, till he finally became nearly impotent, but afterwards recovered under proper treatment.

Many such cases are witnessed in persons affected with paralysis, and other diseases, which, with the proper treatment, almost invariably recover. But in cases occurring from accidents as mentioned in the above cases, the cure is very uncertain, especially if they do not receive the attention proper and necessary without too much delay, for if the organs have time to become withered, the vessels do not secrete, and consequently become closed. In such a case, recovery is impossible, for the treatment cannot reach the affected part.

This brings us to the conclusion, that the strength of the sexual desires depends in a great measure upon the organs themselves; yet there are other influences that may operate upon them, as we have previously shown, according to the various mysterious sympathies emanating from other parts. The celebrated John Hunter gives us an instance he met with in his own practice. I have had many like it in mine. The patient was perfectly incompetent for the marital duties, solely from fear.

CHAP. IV.

SPERMATORRHŒA, OR THE UNNATURAL AND EXCESSIVE LOSS OF SEMEN.

When this is lost at the commencement of masturbation, it is supposed to be entirely harmless, because they think by such a practice, they will escape, to them, the immoral and degrading vice of licentiousness, and therefore by the practice of it, when their ardent, youthful and healthy desires are too overpowering, they think they are doing no more than fulfilling the behests of nature. In this idea lies the great danger, not only to themselves, but their companions. This is the cause of its being so very prevalent and destructive, for before the first one of a circle of acquaintances is warned of his error, by its effects on his former vivacity and weakness, nervousness and probably his own discovery of its preventing the proper development of youth, hundreds may be in the same path to ruin, by its baneful effects on the mind, and the consequent premature decay and death of its victim. The progress of course is gradual—first the vivacity and energy of youth is changed to listless indifference; the vigor of manhood remains no longer; the peaceful content of mature life is changed to despondency and gloom. I have had many a patient, whose mental powers at the opening of the season of life were capable of making him (what every intellectual person desires,) both eminent and happy, but whose bodily health and strength being lost, the mind becomes shattered and almost annihilated in this vortex of destruction, tempts the sufferer to end this short lived, but miserable existence by suicide.

The loss of semen by Spermatorrhœa affects the individual very differently from excessive indulgence, for the loss in every instance by the former, is an operation of, or caused by the mind, whereas, that of the latter, is, even in the greatest excess, from the effects of a locality of feelings; yet grave authors, pretending to know all that can be learned of these complaints, say the effects are much the same. It is only necessary to convince the reader of the absurdity of their assertion, by a moment's reflection, or, may be, to an acquaintance who is notoriously licentious, but still as far as appearances are concerned, enjoying good health. Not, so however, with the Onanist, for a partial acquaintance even will discover the inroad that is being made on his system. Do not understand me as saying, that a licentious career does not injure health; far from it—but the effect is not the same as with the Onanist.

An author of a work on this, and the various sexual diseases, divides this complaint into three different stages: "The first stage," he says, "is when the complaint is in its incipient stage, and its effects are entirely local: The second stage is when the bladder as well as the surrounding viscera are involved: Third, the mind as well as the whole system is involved." Now I am surprised that any surgeon who has had the advantages of the practice in these diseases which this gentleman, from the number of years he has devoted to them, must have had, will assert that these emissions are wholly a local complaint. By comparing the seminal losses with venesection, the reader can form a faint idea whether this is a local affection or not, for, from an investigation of the subject, a loss of one ounce of the former is equal to a loss of forty ounces of the latter.

Those persons who can see that the semen is lost are, generally speaking, only those whose cases are known. But, at least one half who suffer from this loss, are unconscious of the fact, as it escapes with the urine, while at stool, or by gradual leakage, and these are the ones who suffer most, for in nearly every case they are under the care of physicians, who grope in the dark, and try, from a want of true knowledge of the case, to cure the effect without removing the cause. While thus laboring, they will treat them for liver-complaint, disease of the heart, indigestion, or symptoms of almost every disease the system is subject to. The ravages of this destructive disease are not confined to a particular class, to age or condition, nor, as I think, I have satisfactorily shown, is it always a consequence of vicious conduct, as some suppose, but, on the contrary, any severe attack of sickness often causes it, so the most exemplary and virtuous may be the victim, for the causes of it operate on the healthy and strong, as on the sickly and weak, and attack the middle-aged, either married or single, of the most temperate habits. Therefore the incalculable misery—unless this destructive pestilence is unveiled—that every one may know how to guard against its evils.

As I have before stated, the genital organs are so intimately connected with the rest of the system, that the slightest derangement of them affects it, more particularly as some parts are in common with the Urinary; the Rectum also is in close juxtaposition to them, so that any disease that affects the genital organs is very apt to disarrange all these likewise, and it is very probable this second affair may be more severe than the first one

Diseases of the bladder, kidneys, urethra and rectum, are quite common, and often very distressing, while the causes of them remain undiscovered. In many cases these diseases are only symptoms—the primary trouble being Spermatorrhœa. To this extent is this the fact, that five out of every six of my patients have a disease of the bladder to that extent as to have worn pads, braces, &c.; but as the cause is not known, of course the effect is not removed.

The main cause of the genital organs exerting such an influence over the whole system, is their intimate connection with the nerves. As there is no process carried on in the body requiring so much the action of the nerves, as the evacuation of the semen; that is the reason it is so exhaustive of the vital energy. In youth this power is brought in requisition, for the formation and full development of the genitals; so if nature is not left free for this purpose, the other parts of the system are not stimulated in a sufficient manner for the perfection intended, and necessary for a full development of the entire system. This is the reason why those who are castrated are always imperfect in body and mind, and descend to the grave so early, for if these organs are removed, nature will not sufficiently concentrate itself to any one so as to be sufficient to stimulate the whole; consequently, the development is only partial. Eunuchs are a sufficient proof; or compare a mutilated animal with one that is not, and it will be at once evident that the form of the body and the disposition are completely changed. Even in after life, the vital energy requisite to secrete properly, at the same time invigorates the system, and disposes it to a constant activity, that otherwise would not be exhibited. Therefore, this substance is an essential and important agent, not only for perfecting the body in early life, but at a later period to arouse it to the proper exertion. So, I think, there can be no doubt that a deficiency or an abundance of this fluid does exert a most decided influence on the character of the individual. Many are dull and inactive, from a deficiency in this respect, while others are too impulsive and restless, from a superabundance—yet the cause is seldom suspected. And if this nervous energy is exhausted by emissions, its abstraction necessarily weakens the power of the whole system. By an extra effort nature can, for a time, supply the deficiency produced, but if the excess be continued, this effort soon fails, and then the general prostration follows. Then it is the stomach cannot digest the food, the heart cannot, for the lack of power, propel the blood, nor the brain think.

This brings us to the cause, in such cases, of dyspepsia, heart diseases, fettered intellect, insanity, and the thousand and one other evils. It has been stated that the brain has been found wasted and softened, after the death of persons from these complaints, and I have no doubt that these poor victims, who are troubled with pain in the head, dimness of vision, loss of hearing, &c., the brain is constantly in a state of inflammation, or wasting away, and continues till they become deaf, blind and insane.

CHAPTER V.

CAUSE OF SPERMATORRHŒA

The most frequent cause of Spermatorrhœa, of course, is the degrading habit of self-abuse ; yet a too frequent sexual excitement will cause it. Another cause which will produce these emissions is an innocent habit, which has, up to this time, never been mentioned by any writer. Many a person has been convinced that this disease has been caused by some excess or other ; but what that excess was, they were unable to tell ; but they attributed it, generally, to a too full indulgence of their sexual passions, either at an early or late period. One thing was certain—they had all the misfortune of the masturbator, without the crime of having knowingly produced them.

Some may have become impotent—partially or wholly —and some may have only had the other symptoms accompanying Spermatorrhœa. The habit to which I allude, is that of only a partial connection ; that is, withdrawing the penis, just as the semen is about to be ejected, in the act of copulation. The reason for this is, to prevent the female becoming pregnant. Why this practice is injurious to the male is, because the act of copulation, not having been fully completed, leaves the organs to complete the act alone, instead of having the natural stimulus and excitant to the end, thereby leaving more or less semen in the canal and ducts to ooze out afterwards.

An attack of fever and some other diseases will cause Spermatorrhœa, and especially, if the genital organs become inflamed, for it leaves them in a state of irritation, which constantly stimulates them to activity, so that they become entirely independent of the will, for the pleasurable sensations attending the emissions, during sleep, soon pass away, and then they begin to occur, without the sensation or knowledge, till awakened in the morning; and finally, they take place in the day-time, when the bowels are moved, or the urine passed; but in the most aggravated cases there is a constant running away of the semen, without any intermission. This is from a relaxation of the mouths of the seminal ducts. When it comes away with the urine, it is from an inflammation of them, which involves the bladder, and causes a desire oftener to pass urine than usual.

I very frequently have married men for patients, both from self-abuse at an earlier age, and which they supposed the marriage state would rid them of, which is a fatal idea, for it involves the misery of an unsuspecting person as well as their own. It is often the case, that Spermatorrhœa is caused by sexual excesses, and not using true moderation. This is a fatal error; for they are liable to suffer at any time. And they are greatly surprised, when they find on coming to me, what the real trouble was. It is the same with those moral and religious young men, who have never been addicted to the practice of masturbation, yet the mature development and good health, will cause these dreaded emissions at night, and with the consciousness of strict rectitude, they think it is a harmless occurrence, until the fatal error is found out too late for them to escape its ravages, for I can assert positively from the advantages of an immense practice, that an emission in this unnatural manner is injurious. This is an important truth, and should be known by every one; and it, no doubt will surprise the reader, because it has not before been known. For that very object I have published this little work, divested of all indelicate language, that it may be read in every family without any objection from the pure-minded female. This insuperable objection to all the other works, is the cause of so much ignorance, which for the benefit of humanity should be supplanted by knowledge.

There are certain medicines that will cause this complaint, and often cause severe inflammatory complications, and yet they are administered to prevent the difficulty; such is especially the case with Canthaeides, Phosphorus, Iron, Opium, &c.

CHAPTER VI.

Effects of Spermatorrhœ—As shown by cases from my own note-book, to all of which are attached their certificates to me, of their cure. The originals can be seen at my office with the full names. And I am not afraid to assert, that they are unprecedented.

I will preface these cases by enumerating the effects of the disease, as will be seen, by reading the cases of the different individuals.

Some of the first effects, are exhibited on the genitals, particularly those most immediately connected, such as the urinary organs. The irritation speedily extends from the ducts and vas deferens to the urethra and bladder, and when it is so great in the latter that it will not retain the least quantity of urine, it proceeds along to the ureters and kidneys, causing all the symptoms of inflammation of them, and of gravel, with great weakness, and pain in the back. Many patients come to me for their cure, without any idea of the real cause of all their suffering, for nearly all of them have been under the charge of physicians, who have tried in vain to alleviate their condition in the least. The rectum, or large

PLATE 21.

Fig. 54.

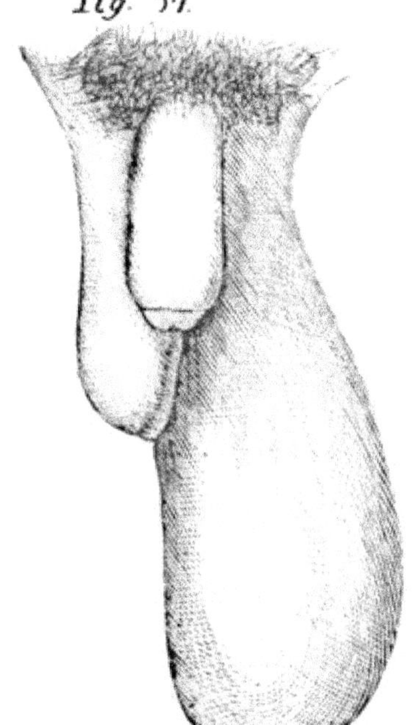

Fig. 55.

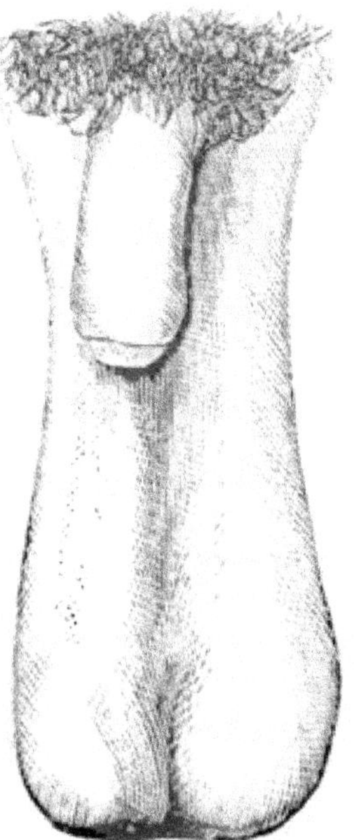

intestine is very apt to suffer, as it is in direct proximity to the prostate gland and seminal vesicles. I have had patients, with a sensation of bearing down to that degree, that they have worn trusses and bandages, by the advice of physicians, to help support the bowels. The constant irritation and inflammation internally, soon causes constipation and piles; and when the mucous coat is involved, then it causes a diarrhœa, which no medication can check, so long as the cause is not removed. A frequent symptom is a peculiar irritation or itching of the urethra and meatus-urinarious, or opening of the canal, sometimes little sores or pimples filled with water or blood will appear on the external parts. These local effects will soon usher in those more severe—universal lassitude and weakness, with a remarkable loss of power in the legs, causing great weariness in walking, and especially in ascending stairs, which will cause trembling, paleness, a fluttering or palpitation of the heart, and fainting indications. In time the stomach suffers, digestion is imperfect; emaciation, and wasting away follows.

The loss of the nervous power affects the diaphragm; and then there is a difficulty of breathing, which finally affects the lungs. The head is not exempt from the general influence, and headache, rush of blood to the head, dizziness and drowsiness, dimness or loss of sight, is frequently observed, as if a cobweb was spread over the eye, which will fill with water, and look red on any extra use of it; generally, however, the lids are more affected, which it is impossible to benefit until the cause is removed. The effect on the mind and feelings are the most marked. Mental activity is as difficult and unpleasant as bodily. They become dull, listless and moping; the memory fails and the judgment weakens, so that all power of application is lost. The mind cannot be concentrated; when an effort is made they will wake as from a dream, and find they have quite forgotten the subject. This, of course, renders them entirely unfit for business, or any close application, and finally degenerates into insanity. If the records of the failures in business, or in the seminaries, academies, and colleges, could all be written, the numbers that were made victims from this disease would astonish the world.

Moral teachings in these cases only add fuel to the

flames. The feelings and dispositions of this class of persons are almost indescribable. They become melancholy and sensitive to such a degree, that they will often shed tears at a trifling occurrence. They imagine themselves subjected to trials and insults that no one else would dream of. They will start suddenly at a slight noise. What makes it worse then, is an aversion to the opposite sex. Many of the misanthropes and hermits are of this kind.

I will now pass from the more general description to the special ones, as presented by the cases themselves.

CASE I.

In 1850, Mr. S., of New Jersey, gave me this written statement:—

"I am 38 years of age. I commenced the habit of masturbation at the age of seven or eight. I was taught the practice by a female servant, and continued it with more or less violence, until I was married, at the age of twenty-seven. I several times previous to marriage, tried to leave off the practice, but found by so doing, I had involuntary emissions at night; to avoid which, I continued the practice of it till I was married, hoping that would bring me all right; but how sad were my feelings, and how miserable my position, when I found my hopes were blasted. Well, in my endeavors to prevent the emissions, my habits led me into excessive venery, which, in my then state of health, was nearly as destructive, for I unfortunately continued it for nearly six years, until I became so reduced, and in such a dreadful state of nervousness, that I tried to abstain in toto, in order to recover my strength; but in doing so, the horrible emissions again commenced. I then tried to refrain gradually; but that would not do. My situation was truly horrible, for I was descending to the grave at a rapid rate. I then applied to a distinguished surgeon for a cure; but his greatest efforts were of no benefit. I then tried another, with no better success; then another, and another, with the like result. I then called on a surgeon, who devoted his attention entirely to this class of diseases, and who charged me a fee of $200. I will state further, that you can fully understand my case. I am a complete wreck—my mind has long since lost its vigor, which threw my business into such a deranged state, that I became involved to the extent that extrication was impossible—and I lost every dollar. I am now scarcely able to fill a clerkship, upon a small salary. I am very nervous, can eat only the most easily-digested food, am troubled with costiveness, short breath, and so weak that I can scarcely ful fill my duties. I have been using the cold-baths daily, for a long time. I have tried salt-water baths, and, in fact, everything that I have been told or supposed would be a benefit to me. Now, Dear Doctor, after hearing of some cases that some acquaintances and others had experienced under your treatment, I place myself in your hands. And may God assist you in restoring me—for this is my last and only hope."

This gentleman's case is a simple narrative in his own hand-writing. His sleep was broken and restless; he at times complained of vertigo, noise in the head, loss of sight and memory; he also had urethral irritations, attended with an occasional discharge, pains in his loins, and groins. Notwithstanding the early commencement of the habit, and its long continuance, and of his excesses after marriage, I was after a persevering treatment of some six weeks, able to recover all of the general derangements of the system—except entirely restoring his digestive powers fully, in so short a time—and prevent the emissions entirely, except one, which took place within a few days of the commencement of my treatment.

This case is a fair copy of the results of those who are so unfortunate as to follow the advice of those physicians who advise marriage, when a person is laboring under a disease of this nature. It surprised me much, to find the organs so strong as they were, for it is almost always the case with these married patients, that they are nearly or partially impotent.

CASE V.

A case resembling this very much, was Mr. M., of this city, who I cured about a year ago; but he had been entirely impotent for four years. He had not been able to discover any semen for the whole time. On my examining the urine, however, with my microscope, I discovered some in a very vitiated state mixed with it, although with the naked eye, it was not discernible. This gentleman was some ten years younger than the other, and had been under the charge of a number of physicians, without improving his condition in the least. They finally prevailed on him to get married; he took their advice, not only to his own sorrow, but one who was dearer to him than life, in his then unhappy condition, for he had deceived an over confiding and innocent wife, almost as soon as they were pronounced man and wife, for scarce six months had intervened before he found himself in the deplorable condition mentioned above.

Now, this one case I know, by being placed before the public, will be the means of saving thousands from a similar fate, for certainly no sane person, who has ever been addicted to any practice from which any such horrible contingency could possibly arise, will even think of taking so fatal a step, without consulting me or some other physician in whom they can trust themselves, with the conviction that their case cannot be in better hands.

CASE III.

The next case is that of Mr. P., of this city. He is a young man of high intellectual power, and general business talents—a junior partner in a wholesale house. A few months previous to his placing himself under my charge, he had a very severe attack of gonorrhœa, with orchitis, which left a stricture, and of so inflammatory a character, that he was obliged to pass urine frequently. His health otherwise was tolerably good. These were his symptoms at the time of his coming to me With the addition of his having been impotent for some time, with a very great relaxation of the scrotum, which allowed the left testicle to descend as much as six inches. I had no trouble in restoring him to health, in a short time, with the single exception of the last above mentioned relaxation of the seminiferous tubes. In some of these cases, it is sometimes necessary to operate, in order to bring the scrotum back to its natural size and firmness. I did not do so, however, in this case ; but by his wearing a suspensory bandage, and using the washes I gave him, succeeded in nearly restoring it to its former firmness. I consider this diseased state of the testicle one of the most dangerous and intractable of all these complaints, for if so bad as to require an operation with the knife, there is great danger of the virile powers being destroyed.

I will now copy a case of M. Lallemand's—a celebrated French practitioner, who has, from his assiduous application to this subject, been the means of enlightening the world to a greater extent than any other person, of the importance of understanding the treatment of this dire disease. His treatment, however, I will mention here, is one that I do not approve, and of course don't adopt, as far, I mean, as cauterization is concerned; yet, he thinks, and so asserts, that out of the whole number of these complaints, there are two-thirds of them that cannot be cured, unless that treatment is adopted. In this he is in error, very greatly so, as will be seen in the cases of my own, in the succeeding pages. I have not used such treatment in scarcely a case ; yet the certificates for their cure I have in their own handwriting, as I have previously stated that all the cases I publish are those for which I have their own certificates, and some of them had, previous to applying to me, been treated in that way. Some even had submitted to cauterization, eight or ten times, and left worse than when they commenced. Yet the physicians adopt that treatment in nearly every case. This case confirms what I have said about married men suffering with the disease as well as those who are single, and shows how easily physicians mistake the symptoms for those of other affections.

CASE IV.

He (Lallemand) says, "In the month of January, 1824, I was requested to see M. De S——, who was affected with symptoms of cerebral congestion, from which he had suffered for some time. During several consultations I gathered the following facts :—

"M. De S. was born in Switzerland, of healthy parents, and his father died suddenly of affection of the brain. M. De S. possessing a strong constitution and an active mind, received an excellent education, and at an early age turned his attention to the study of philosophy and metaphysics; he afterwards studied moral philosophy, and politics. After having spent some years in Paris, pursuing his favorite subjects, he was obliged to undertake the management of a manufactory, and to attend to details, which wounded his pride. He became, by degrees, peevish and capricious, was irritated by the slightest contradiction—showed no pleasure at fortunate events—and gave way to anger on improper occasions. At length he appeared to feel disgust and fatigue at correspondence or mental exertion. At this period he married, and Dr. Butini, of Geneva, his medical attendant and friend, wrote respecting him, as follows:—

"'With this marriage, the most happy period of his existence seemed to commence; but soon the germs of the disease, which so many causes had contributed to produce, became rapidly developed. It was perceived that M. De S. wrote slowly, and with difficulty, and his style presented signs of the decay of his faculties; he stammered and expressed his ideas very imperfectly; he experienced also, at times, attacks of vertigo, so severe as to make him fall, without, however, losing sensibility or being attacked with convulsions. One day, an attack which frightened the patient seriously, and left a deep impression on his family, came on whilst writing an ordinary letter. His medical attendants attributed this attack, which left a weakness of the right side of the body, to apoplexy. Twenty leeches were applied to the anus, and the danger seemed at an end. Similar attacks, however, occurred at Geneva and Montpelier, and several distinguished practitioners were consulted. Some of them, struck by the misanthropic irritability of the patient, and his solitary habits, regarded the affection as purely hypochondriacal, or nervous; others, taking into consideration his digestive disorder, considered it an affection of the brain, such as encephalitis, or chronic meningitis, arising from hereditary predisposition. This last opinion was held by Dr. Bailly, of Blois. At all these consultations, the necessity of abstaining from serious occupation, the utility of traveling, of various amusements, and of a strict regimen; and the importance of free evacuations from the bowels, by means of purgatives and injections, were agreed on. Many of the practitioners recommended the frequent application of leeches to the anus, with milk diet, &c.; others thought, that assafœdita baths, and camphor, were indicated. None of their modes of treatment, however, produced any considerable amendment; the leeches weakened the patient, and the milk diet disordered his stomach; his constipation continued, cold plunge-baths and cold effusions to the head relieved the insupportable spasms. M. De S. experienced great pains in his legs and face; the waters of Aix, in Savoy, and the use of douches also appeared to produce some improvement. Still M. De S. became more irritable, and at the same time more apathetic. His attacks were more frequent and more violent, and he manifested greater indifference towards the persons and things he had before been partial to. The weakness of his limbs increased to such an extent, that he frequently fell, even on the most level ground. His nights were restless, his sleep very light, and often interrupted by nervous tremors, or acute pains, accompanied with cramp. The cerebral congestion increased, and the imminent fear of apoplexy rendered leeches, to the arms, venesection to the foot, tartar emetic ointment, blisters, mustard, pediluvia, and the application of ice to the head necessary. Notwithstanding the employment of these energetic measures, another violent attack of congestion occurred. I was summoned on this occasion, and I found the patient restless, agitated, and incapable of remaining two minutes in the same place; his face was red, his eyes projected and fixed, his physiognomy expressed extreme dread; his walk was uncertain, his legs bending under the weight of his body, his skin cold, and his pulse small and slow.

"'The last circumstance attracted my attention, and I also recommended the application of leeches to the anus. M. De S. immediately threw himself into a violent passion, and asserted that leeches had always weakened him, without giving him any relief. I was too much afraid of the occurrence of apoplexy to pay attention to this assertion, and I succeeded in obtaining the application of six leeches. The next day I found the patient very pale, and so weak he was unable to walk—a source of much annoyance to him, as he manifested a constant desire for motion. An œdamatous swelling of the parotid gland and of the right cheek followed, which was succeeded, a few days after, by a similar state of the left leg and foot. Sleep had become indispensable, and the patient was much reduced from want of it; he told me, with tears in his eyes, that he had lost his appetite, and could no longer relieve his bowels. I also learned that he was habitually costive and flatulent; that he often had recourse to injections and purgatives, in order to relieve his obstinate constipation; and, lastly, that his walks and the evacuation of his bowels had lately become the sole objects of his thoughts and conversation. Having observed analogous symptoms in almost every person affected by diurnal pollutions, I made further inquiries respecting the attack in which it was supposed the right side had been paralyzed, and I was soon convinced that the intellectual powers had been wanting, and not the power in the hand which held the pen; both sides of the body had, in fact, retained an equal degree of strength.'

"Struck by a remark of Dr. Butini, respecting the progress of the disease soon after marriage, I made inquiries of Madam De S., and learned that the character of her husband had become so uncertain and tormenting, that his friends thought he must be unhappy in his marriage. I then suspected that the origin of the patient's disease had been mistaken; and I requested that his urine might be kept for my inspection. The appearance of the urine was sufficient to convince me that my suspicions were well founded; it was opaque, thick, of a fœtid and nauseous odor, resembling that of water in which anatomical specimens had been macerated. By pouring it off slowly, I obtained a flocculent cloud, like a very thick decoction of barley; a gluey, ropy, greenish matter remained, strongly adherent to the bottom of the vessel, and thick globules of a yellowish, white color, non-adherent, like drops of pus, were mixed with this deposit. I was therefore convinced, that Spermatorrhœa existed, together with chronic inflammation of the prostate gland and suppuration in the kidneys.

"Notwithstanding the state of M. De S's intellect, I was able, at a favorable moment, to obtain further information:—At the age of sixteen he had contracted blennorrhagia; this he carefully concealed, and succeeded in curing by the use of refrigerant drinks. The following year the blennorrhagia returned, and was removed by astringents. Two years afterwards, from drinking freely of beer while heated, the discharge again appeared, and after some time it again returned, from the effects of horse-exercise. Since that time M. De S. had felt little sexual desire, and had abstained from intercourse without regret. Ejaculation during coitus had always been very rapid.

"Fully convinced by combining all these circumstances, I explained to M. De S. the nature of his disease, and he promised me to observe carefully. The next day he called me aside, and told me that the last drops of urine were viscid, and that during an evacuation of the bowels, he had passed a sufficient quantity of a similar matter to fill the palm of his hand.

"Eight days after, another attack of cerebral congestion occurred, followed by stentorous breathing, cold skin, and an inappreciable pulse; the patient fell into a kind of syncope, of which he died on the 1st of March, 1824."

Every few weeks I have cases as nearly as possible to the above, and find it is the fault of the treatment used for the cure of the gonorrhœa—using astringent or caustic injections and such medicines as copaiba, cubebs, &c. Yet this treatment is still universally followed. By my own treatment in these cases, I find no trouble in curing them—in almost every instance, within twenty-four hours—whether they are of recent or of years' standing; and I have never lost a patient, notwithstanding many were of the most horrible description of constitutional venereal affections.

Constipation is occasionally the cause of Spermatorrhœa, as I have before stated; but very few would suppose, however, that it could have the effect which sometimes follows from it.

The next Case will convince the most skeptical—though it is seldom we have so severe a case—they are generally pretty bad.

CASE V.

"M. De B. consulted me, (says M. Lallemand), in the month of May, 1834, respecting a cerebral affection, on whose nature distinguished physicians could not agree, but which all regarded as very serious.

"He was of middle height, with a large chest, and a well-developed muscular system, his hair brown and curly, his beard thick, his face full and deeply colored. Notwithstanding these signs of apparent strength and health, I noticed that his knees were slightly bent, and that he was unable to remain long standing, without shifting the weight of his body from one leg to the other; his voice was weak and husky, the motions of his tongue seemed embarrassed, and he articulated his words in a confused manner; his attitude was timid, and his manner had something of incertitude and fear; he had been married fifteen days. His mother-in-law and his young wife, who accompanied him, informed me, that within this period he had had several attacks of congestion of the brain, during which his face was highly injected. At the first of these attacks, the surgeon, called in the night, had bled him to the extent of three pounds, in order to prevent apoplexy; repeated venesection, and the frequent application of leeches had relieved such attacks of congestion, but had not prevented their recurrence. The patient had become subject to attacks of vertigo, and was unable to look upwards without feeling giddy; his legs had become so weak that he had fallen several times, even when walking on level ground. His ideas had lost their clearness, and his memory failed rapidly.

"These symptoms had spread consternation through both the family of my patient, and that of his wife, especially as several practitioners of reputation were agreed as to the existence of some serious disease of the brain, although they could not decide as to its nature. Most of them, however, were inclined to suspect ramollissement (softening). The countenance of the patient, during this recital, the coincidence of the congestion, with the period of his marriage, and the bad effects of blood-letting, made me suspect the nature of the disorder, and induced me to question the patient, separately.

"When we were alone he told me, stammering, that an unexpected occurrence, immediately after his marriage, had at first prevented any conjugal intimacy, and that afterwards he had found himself completely impotent. He attributed this misfortune to the attacks of cerebral congestion, and to the bleedings he had undergone. On further inquiry, however, I discovered that he was affected by diurnal pollutions.

"The following is the history I obtained from this patient, by dint of questioning:—At the age of sixteen, he possessed a very strong constitution and ardent and passionate character. At school he contracted the habit of masturbation, and at the end of three months he had frequent noctural pollutions, with pain in the chest, and troublesome palpitations, which warned him of the danger of the vice, and he renounced it forever. When he became free from the restraints of school, he subdued the ardor of his temperament, by the most violent exercises—especially that of the chase—and he attached himself to agricultural pursuits, with much energy. This new mode of life so completely reëstablished his health, that he was tormented by energetic and continual erections, to subdue which he employed river baths, even in the coldest seasons. He never committed excesses of any kind, and had never suffered from any blennorrhagic or syphilitic affection. In 1831, the erections were slightly mitigated, but he became very much constipated, which he attributed to the constant use of horse-exercise. In 1832, he experienced some numbness and creeping sensations in his feet and legs. In 1833, frequent dazzling of sight occurred, with vertigo, difficulty of vision, and flushes of heat towards the head and face. The patient attributed all these symptoms to the effects of his still increasing constipation. At the same time that these symptoms occurred, the patient's erections became rarer, less energetic, and, after a time, incomplete; his fitness for intellectual labor diminished; the cerebral congestions became more frequent and more severe; his face became habitually very red, his head burning, and almost constant fixed pain came on in the orbits, and his character became fickle and contradictory.

"His family physician, attributing all these disorders

to a state of plethora, caused blood to be drawn severa. times, without benefit.

"In March, 1834, M. De B. engaged himself to a young lady, who lived about two leagues from his estate; and in order to visit her without neglecting the care of his property, he was obliged to make long and frequent journeys on horseback. Shortly before his marriage these journeys became so frequent that he might be said to pass the greater part of his time on horseback. His constipation increased to such a degree that he passed forty days without fœcal evacuation. During his efforts at stool, he passed semen in large quantities, and in jets, although the penis remained flaccid. He had previously several times noticed the same occurrence, but as he attributed it to his long-continued continence, he paid little attention to the circumstance. His urine was constantly muddy; it was passed slowly and with difficulty, and threw down a large quantity of thick and flocculent deposit.

"M. De B. awaited the period of his marriage with a vague uneasiness, of which he could not imagine the cause. He was much attached to his betrothed, but nevertheless, he experienced more embarrassment than pleasure in her society.

"I have already stated what occurred after his marriage. I should add that, having examined his genital organs, I found them, contrary to my expectations, of unusual development; the testicles were large and firm, but the scrotum was slightly relaxed. The patient experienced a strangling in the organs, and at times felt as if they were compressed by a hand of iron. These sensations increased when near his wife, and the penis diminished in size and became retracted towards the pubes, in proportion as he endeavored to excite erection. The union of all these circumstances could not permit any doubt to remain on my mind as to the nature of his disease. It became evident that all idea of cerebral affection must be abandoned, and that the diurnal pollutions, with all the symptoms of which they were the cause, must be referred to the patient's constipation. The first indication to be fulfilled, therefore, was to relieve the constipation; indeed, I hoped that was all that would be necessary. The youth of the patient, the development of his genital organs, and the strength of his constitution induced me to suppose that his cure was prompt and easy. Things did not, however, follow so simple a course. The next day, the patient began to use ascending douches; and was put on a vegetable diet, with iced-milk.

"The first douches caused the evacuation of an immense quantity of fœcal matter in lumps, as hard as bullets, and it was not till after the sixth douche that the fœces were of normal consistence. I then caused the temperature of the water to be lowered to about 85° of Fahrenheit, and afterwards to 81°. The last few douches were given at 68° Fahrenheit. After the twelfth douche had been administered, they were omitted, the bowels having acted regularly every day, without the necessity of the slightest straining. By this time the patient's countenance had lost its purple tint, and presented a more natural appearance; the straining sensations of which he had complained, diminished by degrees, and at length disappeared entirely; his legs regained their strength, and he was able to continue in a standing position for a long time without fatigue, and to take long walks without inconvenience; his voice resumed its natural tone; his eye regained its expression, and all his motions acquired firmness. At the expiration of a fortnight the spermatic discharges during defæcation had ceased entirely; but his urine still continued thick His erections had already acquired sufficient energy to make him believe himself cured; but ejaculation took place almost immediately. The use of ice and cold lotions, did not ameliorate his condition.

"Such was M. De B's state, at the end of a month, when, in order to act directly on the orifices of the ejaculatory ducts, I determined to cauterize the prostatic portion of the urethra As soon as the inflammation had subsided, his erections became more perfect and energetic, yet ejaculation still took place too rapidly

"The period for using the mineral water having arrived, I sent M. De B. to Aix, in Savoy, where I visited him shortly after. He had experienced very little benefit from the use of the waters, either externally or internally

"I now prescribed douches, alternately very warm and very cold, on the perineum and loins, the spout being changed when the sensation of either cold or heat became very intense. The bath was ended after about twenty or twenty-five minutes, by the cold douche, and the patient's skin remained highly injected for some hours afterwards. The effects of these douches were conclusive. After the first, the patient's erections acquired a degree of vigor and duration which reminded him of his early torments. He continued the use of the douches for some time after his re-establishment; and when he left Aix, the functions of his genital organs were perfect. Ejaculation was a good deal protracted by the use of the douches.

"I have entered into a somewhat lengthy detail of this case, because the subject affects gravely the most serious interests of society as well as the happiness and peace of families. Besides, I confess that I was much interested by the unhappy position of a young man whose misfortune was undeserved, and could not have been forseen, as well as by that of his wife—a young woman, scarcely of age, who was obliged to enter into the most unpleasant details.

"It is evident that in the case of M. De B. the constipation was the cause of the involuntary seminal discharges. The patient had practiced masturbation, it is true, and nocturnal emissions followed; but he had continued the vice only three months, and his health, though disordered for a short time, was soon re-established by the use of violent exercise.

"M. De B. was even tormented for several years by erections, which must have been very energetic, if we judge by the means he took to subdue them. From this time he had never committed any kind of excess, and he had never suffered from either syphilis or blennorrhagia. There is then no circumstance in the history of his life, except his constipation, which would account for the involuntary discharges.

"But to what is this constipation to be referred? After all I could learn from the patient concerning his mode of life, I could only refer it to his constant horse exercise. In fact, M De B. sometimes passed whole days on horseback, either for the purpose of hunting or of superintending the management of his property. Shortly before his marriage, his rides became more frequent and longer, and his bowels at this time did not act during forty days. The weakness of his legs, the stunning sensations, &c., increased in proportion as his costiveness became more confirmed.

"This case recalls to my mind the well-known observation of Hippocrates on the impotence of the Scythians, and I have no doubt that his opinion was founded on analogous facts. I shall treat this subject more fully in another place; but since at present I am considering the causes of Spermatorrhœa which act on the seminal vesicles through the influence of the rectum, I report this striking case, showing the effects of long-continued horse-exercise.

"M. De B. was accustomed to nutritious food, and of a well-marked sanguineous temperament; he had a large chest, powerful muscles, and a highly-injected countenance; it is therefore by no means extraordinary that he should have been bled frequently for the relief of the cerebral congestions to which he was subject. On the night of his marriage the blood rushed to his head with greater force than ever, so that an attack of apoplexy was much feared. The weakness of the legs, the frequent falls and the attacks of vertigo, were therefore afterward attributed to an advanced stage of disease of the brain. This was a very natural opinion, but it was an incorrect one; I doubted it from the commencement, although the patient was brought to me in consequence of a supposed cerebral affection. I formed a different impression, because I had previously seen many analogous cases.

"There existed in all these patients something peculiar in the expression of the eyes, in the position, in the voice, and in the general appearance; something of timidity and bashfulness, which I am unable to express, but which is instantly recognized by the experienced, although perhaps it is incapable of explanation. However this may be, the relation of the above case should draw attention to the subject.

"I admit that venesections seemed to be clearly indicated in the case of M. De B., but the loss of blood never produced good effects, either immediate or remote;

and by analyzing the case carefully, his attendants would have seen that under this treatment the attacks increased in frequency. But pre-convictions throw a thick veil over the most acute perceptions. The ascending douches put an end to the constipation; but freedom of fœcal evacuation did not suffice to cure the disease. The seminal discharges, during the passage of fœces, diminished, indeed, or perhaps entirely ceased, but the patient's urine remained thick and muddy, and his erections were incomplete. The application of ice and of the nitrate of silver, and the use of sulphurous waters, were not sufficient to effect this cure; yet there could not have existed any organic change in his genital organs. We can, therefore, only attribute the continuance of the seminal discharge, during the emptying of the bladder, to relaxation of the ejaculatory canals, produced by their long habit of allowing the semen to escape in a passive manner—showing how necessary it is to put an end to the habit as early as possible."

Every patient after reading this case, will at once see the injury which is caused by using the mechanical instruments that are advertised for sale, and warranted to be the only sure cure of the emissions, by preventing them from taking place. Such rings, by being placed over the penis while it is in its natural state, of course becomes tight and causes pain, when an erection takes place, thereby waking the patient; not, however, before the semen leaves the ducts and reaches the canal, which must then either pass out entirely or recede into the bladder, causing all the serious affections heretofore witnessed in the preceding cases, as well as many more, in the after part of the work. I shall also there speak fully on the different appliances that have been tried, as well as that of the medication generally adopted.

The next Case shows the effects of worms in the rectum, not only producing but keeping up Spermatorrhœa, and also in being the original cause of masturbation, even in young children.

CASE VI.

"M. R., a student of medicine, (says Lallemand), enjoyed good health in his childhood, but about the age of fifteen was tormented by prolonged and frequent erections. One evening, for the relief of the itching of which the extremity of the penis was the seat, he rubbed the organ violently between his hands. This led to the establishment of the habit of masturbation as a habit, or rather as a passion, the patient practicing it sometimes as often as eight or ten times a day. His health, by degrees, became so altered that one of his friends suspected his practice, and told him the danger of his situation. By degrees he corrected himself, though not entirely, before he had attained his twentieth year. On his renouncing masturbation, nocturnal emissions supervened, and often occurred two or three times a night. They diminished after a time, but without ceasing entirely, and seminal emissions during defaecation and the emission of urine, were added to them. Thus his health became more and more disordered, for nine years, notwithstanding absolute continence, a severe regimen, and the use of sedatives, tonics, and anti-spasmodies. At length he grew incapable of any mental exertion.

"In 1837, he came to Montpelier, at the age of twenty-nine, in the following condition :—extreme emaciation, face pale, appearance stupid, and confused, intellect dull, reasoning powers much affected, the patient being incapable of connecting two ideas on the most simple topic of conversation ; loss of memory, constant headache, referred to the forehead and temples, and increased by any mental excitement, being then accompanied by nervous tremors, and an almost idiotic state, sleep broken and unrefreshing, constant sighing, frequent attacks of congestion of the head, especially at night, violent noise in the ears, resembling the sound of a waterfall, vertigo, stunning sensations, giving rise to a constant fear of apoplexy, timidity carried to a ridiculous extent, panics of fear, even during the day, character gloomy, taciturn, restless and irritable, horror of the least noise and of all society, irresistible restlessness, great weakness, abundant sweats after every slight exertion, almost constant coryza, frequent dry and hard cough, pains in the base of the chest, the region of the heart and along the spinal column, appetite voracious, dragging at the pit of the stomach, difficult digestion, accompanied with the development of flatus, grinding of the teeth during sleep, burning at the point of the tongue, darting pains in the bowels, especially in the rectum, obstinate constipation, alternating with violent attacks of diarrhœa, stools containing much mucus and sometimes streaked with blood, periodical pains at the margin of the anus, in the perineum, penis and testicles, urine passed in large quantities and very frequently, always throwing down a whitish, thick and very abundant deposit, involuntary emissions during defecation, both when constipated and relaxed, frequent and prolonged erections, by day as well as by night, with constant presence of erratic ideas.

"On sounding this patient, I found the urethra very sensitive, especially towards the neck of the bladder, and I consequently thought that the nocturnal and diurnal pollutions were kept up by a state of irritation, arising from masturbation. I therefore proposed cauterization. This was performed on the following day, and produced the usual immediate effects, but its curative effects did not take place as I had anticipated. I then directed the patient to notice his fœces, and a few days afterwards he told me that he had observed numerous little worms passed in his stools. I now ordered enema of cold water, and salt and water, which, however, produced only a momentary effect—probably because the ascarides inhabited the upper part of the intestine. A few doses of calomel, however, caused them to disappear without returning; and from this moment, the involuntary diurnal emissions ceased entirely, the nocturnal emissions became more and more rare, and the patient's re-establishment progressed very rapidly.

"M. R. returned to his studies with ardor, and long afterwards all functions were perfectly well performed It appears evident that the irritation caused by the ascarides in the rectum, first led this patient to practice masturbation, and afterwards kept up involuntary seminal discharges. I did not discover this at first, because the history of his case, sent me by the patient, was so long, and was characterized by such disorder and want of clearness, that I was unable to arrive at any satisfactory conclusions from such a chaos; his answers were still more vague and unconnected, so that my attention had been chiefly attracted to the state of his intellect and the abuses he had committed. But after seeing the little success of cauterization, and again reading his notes, I paid more attention to the circumstances attending the commencement of his practicing masturbation, and I noticed several symptoms to which I had not before attached importance—such as grinding of the teeth during sleep, burning pain in the point of the tongue, pain in the rectum and at the margin of the anus, the stools always containing mucus and sometimes being streaked with blood, and especially the frequency and durations of the erections, and the constant presence of erotic ideas."

I often have cases, of both sexes, where the original cause of the habit was traced to these little tormentors, and which of course, by the proper treatment, would have prevented all that misery and inroad upon the health of the person, which the result of the habit or loss of semen caused by their irritation, without even having fallen into the practice itself. Their existence in the rectum causes an inflammation, which being contiguous to the seminal vesicles, produce the erotic ideas, as mentioned in the preceding case. Parents should not only be aware of this fact, but the physician should make it his business to inquire particularly if any of these results have supervened; if so, by their being treated at once in a proper manner, all farther difficulties are prevented. This one fact in itself should be sufficient to convince every one, of the absurdity of attributing these complaints to immoral ideas, and that those afflicted would not have been so, if moral suasion had been taught them at the proper time.

There is a fault of the victims themselves equally as great; that is, the desire of keeping their complaint an entire secret. This often is the cause of the most deplorable cases I meet with, for they allow themselves to suffer, because they dislike to consult a physician for such a complaint, or because they have not arrived at that age which enables them to command the money necessary without their parents becoming acquainted with their difficulty, or sometimes even they will consult a physician, and make a false statement of their case, hoping he will be shrewd enough to discover the real truth of the case himself. I often have fathers or mothers apply to me, in behalf of their children, who were so timid that they could not prevail upon them to apply in person. The parents of course became aware of their complaints, by noticing the inroads made on their system or by watching their suspicious actions. I have had many an affecting case presented to me in this manner; but I hope this little publication will meet the eye of those interested in time to prevent many such calamities.

I am satisfied, from the advantages of a large practice, that the tendency to Spermatorrhœa is often hereditary, for I have had some remarkable proofs of it in cases of the same family, and having occurred at the same age. These cases have been so numerous in so many different families, that I have not a single doubt on the subject, for they have not been addicted to the habit of masturbation nor produced by any of the other causes heretofore mentioned. , Congenital pre-dispositions occasionally exist, particularly in those of a nervous disposition, and as very few are aware of the fact, I will give some excellent illlustrations, of these kind of cases, from M. Lallemand's note-book, which fully confirm those in my own case-book; which, though quite numerous, the following will, however, be sufficient to answer all purposes:—

CASE VII.

"In general, such patients were of a sickly constitution and more or less marked nervous temperament; they had been delicate from childhood, and subject to various spasmodic disorders. Some of them presented involuntary twitching of the muscles of the face, hesitation of the speech, &c.; their imagination was active, and their moral and physical sensibility very acute They were very restless, and bore contradiction or mental excitement badly. In childhood they presented local symptoms, which indicated peculiar susceptibility of the urinary organs, every impression of fear or anxiety showing itself in this direction. What would have produced shuddering or palpitation in other children, in them caused a secretion of clear watery urine, which they were obliged to discharge frequently, a sense of constriction of the hypogastrium and a sense of titilation generally accompanied its discharge. This condition of the urinary organs continued more or less severe in all the cases until after puberty, when it became joined with other symptoms.

CASE VIII.

"One of these patients, (continues Lallemand), one day experienced, at the age of sixteen, a fit of irritability and impatience, which, however, he succeeded in repressing; and he then felt a sudden and impetuous desire of micturation; whilst emptying his bladder he perceived a large quantity of pure semen discharged with the last drops of urine. This occurrence was the forerunner of nocturnal and diurnal emissions, which, at the age of twenty-seven, had entirely ruined his health.

CASE IX.

"Another, at the moment of competition for a college prize, was unable to find an expression he wanted; at the same time he felt a want to make water, which he resisted by firmly crossing his legs; but his impatience increased, and he shortly experienced an abundant emission without either erection or pleasure.

CASE X.

"A third patient suffered in the same way, under similar circumstances; he saw the moment approach for sending in his thesis; the more he endeavored to hurry, the less freely his expressions flowed; at length, on hearing the clock strike, he suffered from so great mental disorder that he nearly fainted; at this moment emission took place.

CASE XI.

"A fourth having mounted on a gutter of a high house to take some sparrows' nests, looked down into the court below, and was suddenly seized with such terror that he fainted; on recovering and escaping from his dangerous situation, he found that he had had an abundant seminal pollution.

CASE XII.

"The same circumstances occurred to a fifth, who, in descending a ladder, missed his footing, and fell.

CASE XIII.

"Another patient told me if he looked down from a height, or only fancied himself on the brink of a precipice, he felt a sense of contraction in the genital organs, which passed rapidly to the base of the penis, and ended by causing an emission.

CASE XIV.

"The motion of a swing produced the same effects in a seventh.

"Almost all these excitable persons were exposed to erection, and even pollutions, whenever they rode on horseback. Although all these involuntary discharges were caused by extraordinary circumstances, I should not have paid much attention to them if they had not been followed by nocturnal and diurnal pollutions, which the most trifling circumstance rendered very serious. The disease, however, did not always put on a serious aspect immediately after these singular accidents, very often, indeed, it only injured the patient's health long afterwards; but as its gravity could not be explained by any occasional cause, I feel myself compelled to admit the existence of a congenital increased nervous susceptibility of the genito-urinary organs. Everything indicates, in fact, that the organs of these patients were rather excitable than weak and relaxed; and his condition was congenital, because manifested from the earliest infancy. This excessive sensibility of the genital organs is, however, not always preceded by a similar condition of the urinary apparatus.

"In all these cases, tonics and excitants always produce bad effects; proving the genital organs were not suffering from atony or weakness."

The next Case is one of Dr. McDougall's. It very nearly corresponds with the one from my own note-book, which is the first Case presented.

CASE XV.

"R. H., aged thirty-nine, passed the early part of his life in the country, and was in the habit of taking much and violent exercise. About the age of sixteen, he entered a banking establishment in London, in which by great diligence and steadiness of conduct, he rose before he was twenty-five, to the post of cashier. The affairs of the house fell into disorder, and ultimately a bankruptcy occurred. Mr. H., from the amount of confidence reposed in him by the partners of the firm, was much harassed by these unfortunate proceedings. Soon afterwards he became manager of a large mercantile establishment in the city, and about this time commenced some speculation in foreign bonds. From fluctuations in the share-market, he was a loser to a considerable extent, his mind was much harassed, and he began to suspect those about him of dishonesty towards their employers. On investigation, these suspicions were proved to be totally unfounded.

"Mr. H gave way to great violence of conduct, and resigned his situation. About this time his father died, and Mr. H. was much disappointed at finding that property which he had incorrectly believed entailed, and consequently his, as eldest son, was left by will, to be equally divided between himself and the rest of the family. His conduct at this period was of the strangest description. He dreaded to go out into the streets of the town where his family resided, refused to join in their meals, and ultimately abruptly left their house to return to London.

"In 1837, his state had become such, that in consequence of his repeated letters, members of his family visited London, and on their return, took him with them into Devonshire. About this time, his mental disorder put on a decided aspect; and I had then, as well as later, ample opportunities of observing his conduct, and frequently heard his complaints. Emissaries were constantly on the search for him, to arrest him for unnatural crimes committed in London; every one who met him in the street, read in his countenance the crimes he had committed; tailors made the sleeves of his coats the wrong way of the cloth, in order to brand him with infamy; the sight of a policeman in the street alarmed him beyond measure; and often, if a stranger happened to be walking some little time in the same direction as himself, he would exclaim, that he was one of the emissaries sent to seize him. At other times he would lock himself in his room, and weep by the hour.

"He never took his meals with the family, and never tasted food or drink without first preserving a portion for chemical analysis, as he was convinced his friends were in a conspiracy to poison him slowly, in order to wipe out the memory of his crimes. These ideas haunted him night and day. His digestion was much disordered, his sleep broken and restless, and his bowels excessively constipated.

"His face was flushed, and periodical attacks of cerebral excitement occurred, during which he complained of vertigo, noise in the head, loss of sight, &c. He complained also of loss of memory, and frequently of bodily weakness and lassitude. The best medical advice the neighborhood afforded was obtained, unavailingly; the opinions of the gentlemen consulted were that Mr H. was laboring under aggravated hypochondriasis, complicated with monomania. Various causes were suggested as giving rise to the disorder, but no previous case of insanity was recollected in any branch of the family.

"Mr. H. now began to talk of leaving England, for America, in order to avoid his persecutors; and to prevent this he was placed under the care of a private keeper. While with this person, he frequently and bitterly complained of constant pollutions, while at stool, with darting pain, and a sense of weight between the rectum and bladder. He had also urethral irritation, attended with discharge, pains in the loins and one groin, weakness of his legs, thick urine, piles and costiveness. He kept a diary at this time, which is at present in my hands. Not a day is passed in this diary, without mention of the distressing seminal discharges from which he suffered. These were treated as of no importance by his medical attendants, although he never ceased to complain of them, and solicited aid, so long as he continued in confinement in England.

"When led away from his disorder into any discus sion on public matters, he was, however, a most amusing and instructive companion; as a man of business he was equally acute, and to a stranger, as long as nothing was done to offend him, he was to all appearance, a man of observation and experience in life.

"For about two years and a half he was under the care of various gentlemen, devoted to the insane, and at length he was discharged from an establishment near Bath, by the visiting madistiates, as a person confined without due cause. His first act was to commence legal proceedings against his friends, for his detention, and having gained his action, he immediately proceeded to London, and waylaid and violently assaulted a gentleman of high commercial standing in the city. After this offence, he was confined for a considerable period in default of bail, and immediately on his liberation, it is believed that he proceeded to America.

"From this time nothing was heard of him until September, 1843, when a letter was received by a gentleman who formerly attended him, in which he stated that the same course of persecution was pursued towards him in America as had been followed in England. He complained of not being able to obtain efficient medical treatment, although he had applied to the most eminent practitioners in Cincinnati, and afterwards at Philadelphia and New York.

"After this, nothing more was heard of Mr. H. until the year 1845, when an American newspaper was forwarded to his friends by an unknown hand, containing an account of his death and of an inquest held on him, headed, '*Death of a Hermit in West Jersey.*' It was stated that he had lived on a small farm, entirely alone, with the exception of a dog, and that he had shunned all intercourse with his neighbors. He was taken suddenly ill, applied to a neighboring farmer for assistance, but died in the course of the following day. From information subsequently obtained by his friends, it is believed that he died of apoplexy, or perhaps in one of the attacks of congestion of the brain, from which he frequently suffered before he left his native country."

The symptoms of this unfortunate case strongly resemble those of the thirty-second and fifty-sixth cases related by M. Lallemand, as well as my own. It was more aggravated, however, and presented the somewhat uncommon feature of the patient's discovering the frequent pollutions, and constantly complaining of them; these, unfortunately, were treated as matters of no importance.

Mr. H's insanity, at first, constantly had reference to his having either committed or been accused of committing, unnatural crimes; and this idea never entirely left him, although during the latter part of his life, his many prominent hallucinations had reference to imaginary persecutors constantly watching him, and endeavoring to ruin him, by spreading false reports, and to poison him by adulterating his food and infusing noxious gases into the air. There can be little doubt, on taking into consideration his complaints of weights between the rectum and bladder, with darting pains, &c., in the same region, that the pollutions arose from irritation in the neighborhood of the prostate; and I think that if at an early period of his disease, this had been relieved, there would have been considerable hope of his recovery, from the hallucinations he manifested.

The next case is also one of Dr. McDougall's, and is equally instructive, as it is almost an exact copy of a great many that I have at almost all times under my own charge.

CASE XVI.

"The other case, to which I have alluded as particularly attracting my attention, and which came under my notice about the same time, was that of a young man of high intellectual power and general talents, studying medicine. This gentleman was one of my most constant companions, when, almost suddenly, a serious change came over him. He shunned society, especially that of females; was morose, tacitiurn, and frequqntly shed tears; he sat sometimes for hours in a kind of abstraction, on being aroused from it, could give no explanation of his thoughts and feelings. He constantly expressed to me his conviction that he should never succeed in his profession, and frequently exclaimed that he was ruined both here and hereafter, body and soul, and by his own folly. About twelve months previous to this depression of spirits, he had a severe attack of blennorrhagia, with orchitis and phymosis. This left a degree of irritability in the bladder, which required him to pass urine frequently. His digestion became so disordered that the simplest food would not remain on his stomach, and he had frequent eructations of fluid, which blazed like oil if spit into the fire. This gentleman's father was a physician, and being naturally anxious for his son, obtained for him the advice of many of the most eminent of the faculty. No improvement took place, however.

"After he had been six months in this state, I had an opportunity of spending three weeks by the sea-side, and my friend accompanied me. We slept in the same room, and he was scarcely ever out of my sight. Before our return, his health was almost re-established, and his spirits had returned to their natural condition.

"Twelve months later, however, he again fell into the same state of despondency; and this time his condition was much worse than on the former occasion. He frequently remained in bed three parts of the day, and no threats or entreaties on the part of his father, could induce him to get up. His intellectual faculties were totally prostrated, and a vacant stare, which took the place of his natural lively expression, induced considerable fears of his ultimately becoming idiotic. I was the only person who possessed any influence over him, which may perhaps be attributed to his feeling that I was aware of the cause of his disorder. This state continued between three and four months, during which time I was with him as much as my other duties would permit, and frequently showed him the folly of the course he pursued. At the expiration of this time he gradually recovered. He has since had a slight relapse, once only; he has pursued his professional studies with success, and is at present a medical officer in her Majesty's service.

"On this case, I need not only remark, that the symptoms did not arise from involuntary seminal discharges, but from excessive discharges, caused by abuse. The varied treatment recommended by the distinguished practitioners consulted, proved unsuccessful, because the origin of the disorder was unrecognized, and the remedies consequently useless while the habits of abuse were continued."

CHAPTER VII.

SEVERAL OTHER SYMPTOMS BY WHICH SPERMATORRHŒA MAY BE DETECTED.

In addition to the effects already presented in the foregoing pages, it may be well to note some other symptoms by which they may be known. A portion are frequently observed in other diseases; then again there are others which peculiarly belong to this affection only, therefore enabling us to ascertain their existence, when a more positive proof is not to be obtained.

M. Lallemand says, that "The most abundant nocturnal pollutions are far from being always the most hurtful; yet as they easily lose their character, the habit alone tends to make them more and more frequent. By so doing, they of course are more injurious." He afterwards says, " But this state of excitement is too violent to last long—by degrees the organs become fatigued. Deprived of their natural functions, and, consequently, being unstrengthened by regular exercise, they may at last fall into a state of atony, or the seminal vesicles may preserve the habit of contracting, under the influence of slight or indirect excitement. The evacuations now produce effects quite opposite to those experienced in the beginning. There are, on waking, feelings of discontent, idleness, weight in the head, disorder in the ideas, &c.; but this condition passes off in the course of the day, and the patient is quite well on the following morning, if no further emission takes place. After a time, these effects become more serious and lasting, and two or three days are required to remove them entirely. There is, however, no disease as yet, because the economy is not as yet permanently disordered, but there is a degree of instability in the patient's health, a valetudinary condition, the progress of which it is necessary to arrest."

It will be seen by the above extract that M. Lallemand agrees with me in the very important particular, that the very first involuntary emission should at once warn the patient that the disagreeable sensations produced by it, is the forerunner of the disease itself, being in a rapid course of full development. The dreams themselves, which accompany the emissions in the commencement of these cases, are often erroneously considered as causing them, when, in fact, it is just the reverse—for it is the excitement primarily existing in the genital organs. The symptoms are very much worse when the discharge occurs without being accompanied with the pleasurable dreams, and truly bad when they are disagreeable. The diurnal losses are worse than the nightly ones, especially if they occur spontaneously or from exercize. In very bad cases they occur during the movement of the bowels, especially if they are constipated—or in a very diseased state of the parts, it will be lost in the passing of the urine. Gravel, and other affections of the surrounding organs, will cause sufficient irritation to produce it, and therefore should not be forgotten in searching for the cause. These are the greatest causes of Impotency.

Patients who come to me, are, almost invariably, entirely ignorant of their having lost semen in this way, probably for years, yet when made acquainted with the circumstances, it is generally very easy for them to detect it, whether it is mixed with the urine or lost at the time of defecation. If with the urine there is a sensation of heaviness while passing—sometimes there is a pricking or tingling sensation—the urine looks thicker than usual, and, on close examination, little globules can be seen. The sensation is very similar when the escape occurs at the times of the passing of the fœces, and sometimes a sticky substance can be discovered adhering to the glans. Very often it happens that it flows away, after the movement of the bowels, and the person suddenly discovers it when dressing; a shock then occurs, which is principally felt in the peroneum and at the neck of the bladder. In bad cases it is very difficult to discover the semen in the urine, therefore it should be borne in mind that it never escapes at the first flow, but invariably with the last few drops, or after, as I have before stated, the bladder has been emptied. When there is an ulcerated prostate which discharges itself through the urethra, it would then require an experienced physician to distinguish between the two, as that always comes away last, while in cases of Gleet, or a discharge from an inflammation of the urethra itself, the discharge is driven out, ahead of the urine, and unaccompanied with the sensations previously mentioned. Sometimes patients will have a shaking chill, pain in the nipple, and very often in the back of the neck or in the forehead, shooting pains in different parts of the body, and a heat around the anus. If the disease is not checked, the semen loses its natural appearance entirely, becoming so thin that its character is totally changed

---

CHAPTER VIII.

IMPOTENCY, STERILITY AND INFECUNDITY.

Persons may become impotent from various causes: a long continued gonorrhœa, gleet, stricture or syphilis will produce it as well as sterility, and if the system is so tainted or poisoned with syphilis as to cause the testicles to enlarge and become indurated for any length of time, there is danger of their functions being entirely destroyed. I have had a great many cases of this kind, when the testicles were as heavy and almost as hard as a stone, yet, I am happy to say, I succeeded in every case in bringing them again to their natural state. The emissions destroy the powers of the genital organs, from general weakness, or totally changing and destroying the fecundating properties of the semen. It is a fact, that the semen, when healthy, contains animalcules,

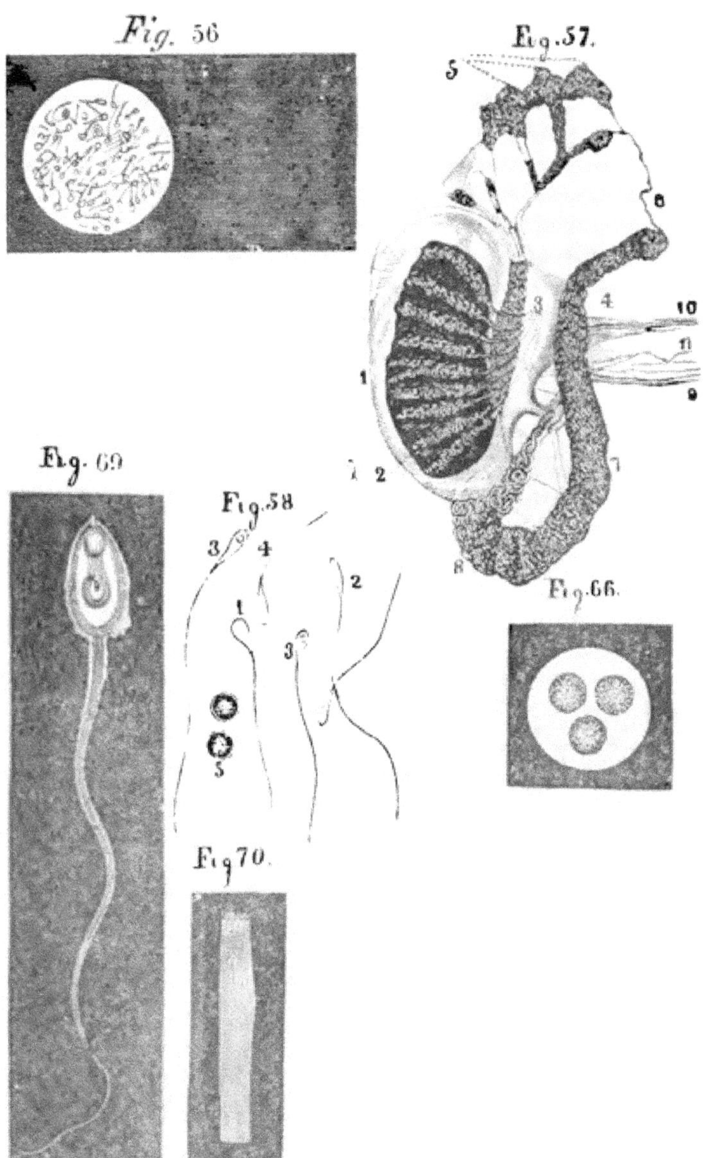
Fig. 56
Fig. 57.
5
6
3
4
10
11
1
9
7
2
8
Fig. 69
Fig. 58
3
4
2
1
3
5
Fig. 66.
Fig 70.

and for them to become fully developed, requires the semen to remain a certain time in the vas deferens and vesicles, after it leaves the testicles, or the zoospermes do not become perfect.

The exact time necessary for the semen to remain in the body, after being formed in the testes, is not known, because it varies in different individuals, and even under different circumstances in the same person. If the organs are healthy at first, they can perfect the semen more rapidly, and lose it often, yet it will remain healthy; but if the excessive loss is continued too long, the parts become weakened, and of course the semen is discharged imperfect. That is the reason why excess, whether from licentiousness or involuntary emissions, is so injurious: for instead of the healthy animalcules which it should contain, there is nothing but the small undeveloped granules, which can, in bad cases, be seen in the urine, like little shining points, or the bran-like grains observable at an earlier period. There are cases so very bad, that the semen is changed to a substance only a little thicker than water; when this happens, it is incapable of stimulating an erection, much less impregnation. This is the reason people become impotent, and finally sterile; therefore, as we know the cause, then the vital necessity of the invalids placing themselves under the care of the proper physician, and I am satisfied this book will be a guide to thousands who have been groping in the dark for years, till hope has almost forsaken them.

I will here inform the reader, that every moment is precious, and no person should be indifferent to the slightest symptoms, and if you are young, so that you are not possessed of pecuniary means, you should at once disclose your situation to your parents, and you will not fail of finding the sympathy you desire.

In a former page I mentioned a vitiated state of the semen that would not stimulate an erection, and that the person would finally become sterile. There are cases of persons, almost fully possessing that power, and yet are sterile; the reason is, that the habits of the body are still vigorous enough to continue the secretion of the semen, but it lacks the health necessary for the perfect formation of the real life—the animalcules. And I attribute many supposed cases of the barrenness in females to this defect in their husbands. As M. Lallemand was the first author who alluded to this important subject, his remarks should be read, at least by those who are interested, that they may fully understand the subject.

"Impotence (says M. Lallemand,) is an absolute cause of infecundity, because it prevents the conditions necessary to fecundation from taking place; but although the act of coitus may be accomplished, it does not follow that the person should always be able to perpetuate his species. Stricture of the urethra may prove an obstacle to the discharge of seminal fluid; or the fluid may be directed towards the bladder or the parietes of the urethra, by deviation of the orifices of the ejaculatory ducts. The secretion may be altered in its nature; it may only contain imperfect spermatozoa, &c. A man may, therefore, be unfruitful without being impotent. On the other hand, I have met with many patients, suffering with diurnal pollutions, who had children exactly resembling them, even during the duration of their disease. Indeed, I have seen several cases in which the disposition to involuntary discharges were hereditary, and they affected both father and son.

"The disease is, however, essentially irregular in its progress. It may continue long, without doing serious injury to the health; long remissions may be experienced or even a perfect cessation of the complaint, for a longer or shorter time. We may easily conceive, therefore, that in the first degree, or during one of the periods of remission, fecundation may take place. When the disease is further advanced, however, many causes occur to render coitus unfruitful. Ejaculation is weak and precipitate, so that the seminal fluid cannot be thrown into the cavity of the uterus; it is not sufficient in order to fecundate, simply to spread the fluid over the vagina; it must be projected with sufficient force to pass through the orifice of the uterine neck. Besides, in these cases the erections, even when they permit sexual intercourse, are incomplete and of very short duration—emission takes place without energy and very soon. This last symptom is almost invariably present in almost all of these cases; and where the parts are weakened, the semen will come away, without scarcely any erection of the penis. But even in the most acute affections, wher the erection is sufficiently lasting to enter the vagina at all, it is only for a moment, before the semen is ejected. Both married and single persons apply to me to free them from this awful situation. The position and feelings of both the male and female at this juncture, can be more easily imagined than described, as this tantalizing of the female often produces the utmost distress. It is, therefore, easily to be seen that during such partial acts, the uterus and fallopian tubes have not sufficient time to experience the excitement necessary to carry the semen to its destination, even when it passes the neck of the uterus. The semen itself also undergoes great changes, to which perhaps the loss of the fecundating power is chiefly attributable. Microscopic researches have elucidated this formerly obscure subject. I have discovered, for instance, that the spermatorrhœa undergo changes similar to those of the fluid which serves as their vehicle; these changes are exceedingly important, and are owing to defective formation. Spermatorrhœa may be met with in a less thick and less opaque fluid than natural, for they are not produced by the same parts or in the same manner; but when the secretion is perfectly thin and watery, the functions are so seriously affected that the animalcules are altered; they are less developed, less opaque, and less active than natural; indeed, they are so transparent that peculiar precautions are necessary in order to make sure of seeing them. Their motions are weak, slow, and cease very soon, and they rapidly undergo decomposition. All these characteristics show how much their texture is relaxed, and how imperfectly they are organized.

"It is evident, the least arrest of development in the spermatorrhœa, must prove an insurmountable obstacle to fecundation, even if the only function of the animalcules be to carry the liquor seminis to the ovum. When, however, their imperfect development only arises from a too rapid formation, it may soon be obviated. It suffices that the involuntary discharges should cease, for a few days only, in consequence of some accidental cause or of one of the spontaneous changes of this extraordinary disease, in order for the desires to become more lively, the erections more energetic and prolonged, and for the functions to be accomplished in a natural manner; fecundation is, therefore, possible, as I have previously stated, during the whole duration of one of these intermissions. This is not the case when the spermatozoa are malformed, rudimentary, more or less deprived of tail, &c., for these changes only take place when there is a serious alteration in the structure of the testicles.

"I have taken every opportunity of dissecting the testicles altered in these cases, and I have always found the secreting structures paler, drier and denser than natural, and the cellular tissue more resisting, and with difficulty allowing the secreting ducts to be separated one from another. Sometimes one-half or two-thirds of the testicle were transformed into a fibrous or fibro-cartilaginous tissue, mixed, in a few cases, with tuberculous matter. It is the same also in the epididymis, when I have even seen traces of ossific deposit in the midst of cartilaginous indurations. These changes, caused by previous inflammation, perfectly explain why the development of the spermatozoa can no longer proceed nominally. Although in such cases the secretion of semen may be more or less diminished, pollutions may still be present if the seminal vesicles have shared the inflammation by which the testicles have been affected, as happens in most cases of orchitis arising from blennorrhagia."

CASE XVII.

"I have, at present, a patient who presents a remarkable example of both these effects arising from this cause; he is now forty-one years of age, and had blennorrhagia, followed by inflammation of both testicles, at twenty-five. Soon after his recovery he married, but has never had children, although the act has been performed regularly if not frequently. He became subject to nocturnal and sometimes to diurnal pollutions, which increased by degrees. His health became disordered, but coitus was still possible. The semen passed, although it presented its characteristic odor, but never showed, under the microscope, other than very small and brilliant globules, without any appearance of tail, but easily distinguishable from globules of mucus, the dimensions of which are five or six times larger. The epididymis of both sides is voluminous and irregular. One testicle is adherent to the skin of the scrotum and the other appears smaller than natural. Malformation of the spermatozoa, therefore, arises from deep-seated changes in the tissues of the testicles, changes which do not permit the animalcules to assume their normal form, and, therefore, render infecundity permanent. To sum up, then—Involuntary seminal discharges may oppose fecundation previously to actually producing impotence, by diminishing the energy of all the phenomena that occur to the accomplishment of the act, and by preventing the complete development of the spermatazoa as well as the elaboration of the fluid which acts as the vehicle for them. These conditions may be rapidly altered by the simple diminution of the involuntary discharges, and fecundation may again become possible. This cannot be the case, when infecundity depends on malformation of the spermatozoa, such malformation arising from permanent alteration in the organs that supply them."

CHAPTER IX.

MICROSCOPIC EXAMINATIONS.

The immense benefits arising from the use of the microscope is of the greatest interest to every one, but more especially is it important to the surgeon. The discoveries by it, more particularly as regards the procreative functions, are of a recent date. M. Lallemand so fully explains the proper mode of pursuing such examinations, that I will below insert his remarks. The process is perfectly simple to those who understand the correct rules, though often requiring a good deal of disagreeable trouble.

"*Microscopic Examinations of the Semen.*—Since the discovery of the spermatozoa, their presence in the seminal fluid has attracted the attention of all who have sought means of distinguishing it from other fluids Microscopic examinations of the spermatozoa, however, not only requires an excellent instrument, but certain precautions, which may be dispensed with in the investigation of coarser objects.

"As the spermatic animalcules can be only seen by means of transmitted light, it is necessary that the glass on which the fluid to be examined is placed, should be of uniform thickness, and without bubbles or striae. The fluid to be examined should be covered by another layer of extremely thin glass made for the purpose, and not by portions of mica, which are seldom free from cracks, and never perfectly transparent. This thin layer of glass is indispensable, in order as much as possible to diminish the thickness of the fluid, to render it perfectly uniform, to hinder evaporation and prevent the object-glass from being soiled by it. A single drop of fluid suffices for a complete observation—a larger quantity always proving inconvenient.

"The little glass that covers the liquid must be firmly pressed down, so as to spread it out, arrest the currents that take place in it, and drive out the air-bubbles. Although the glasses should seem to touch each other, the spermatozoa move with perfect freedom in the space between them, so long as they preserve their energy, and evaporation has not proceeded too far; should such be the case, however, a drop of tepid water favors and much prolongs their motions. However thin the layer of fluid may be, it is impossible to comprehend its whole thickness at once with a very high power; it is, therefore, necessary to alter the focces frequently, in order to be sure that nothing escapes observation. And this is especially important in examining a drop of fluid obtained from diurnal pollutions, because there are frequently only two or three spermatozoa contained in it. It is also necessary to change the position of the reflector frequently, in order to vary the direction and intensity of the light.

"The spermatozoa are often exceedingly transparent in cases of disease, and a very bright perpendicular light is by no means the best for showing them. Varying the density of the fluid under examination, either by adding water or by permitting evaporation, is also often useful. The semen contains matters furnished by the seminal vesicles—the prostate and the urethra—and when the fluid is too thick these matters hide the animalcules. A drop of water applied to the edge of the covering-glass penetrates underneath it and the spermatozoa are more isolated, at the same time that their contour is rendered more defined by the diminution in density of the fluid. On the other hand, the refractive power of the spermatozoa, differs little from that of the fluid in which they are contained, and their thinnest portions are traversed by the light without affording any distinct images to the eye. In this case there are only seen very small ovoid brilliant globules, terminated by a little point. As soon as the water begins to penetrate between the glasses, the rapid motion set up, prevents the objects from being clearly distinguished; but as soon as rest has been re-established the tails of the animalcules appear, and their dimensions seem to have increased in consequence of the diminished density of the surrounding fluid; water suffices to produce this result. It is more sensible, however, when a small quantity of alcohol is added; but the forms of the animalcules are, after a time, altered by this agent; and it is, therefore, advisable to use water only when it is intended to keep the preparation.

"Evaporation sometimes produces not less remarkable changes in the seminal fluid. I have frequently, in cases of spermatorrhœa, failed to perceive anything in the fluid under examination for half an hour, an hour, or more; then suddenly an animalcule has made its appearance; then a dozen, and then perhaps a hundred, in the space of a few minutes. The following morning, when desiccation has become complete, there are no longer any traces of these animalcules, or, at all events, I have been only able to distinguish their tails, the other parts of them being fixed in the dried-up mucus. The absorption of a drop of water has restored the phenomena observed the night before. These phenomena are easily explained; when the refractive power of the spermatozoa is the same as that of the circumambient liquid, the light traverses the whole in the same manner, and the mass appears homogeneous. But evaporation acts more rapidly on the liquid than on the organized bodies contained in it; and when the difference of density alters the refractive power, the forms of the spermatozoa are momentarily defined because they have become more transparent than the remainder of the fluid. When desiccation is complete, however, the animalcules again disappear, because the refractive powers of mucus and dried animalcules are again equal. The absorption of a small quantity of water reproduces the same phenomena, which may be repeated almost indefinitely, since the matter confined between the two layers of glass undergoes no other appreciable alteration.

"In order to be enabled to discover spermatozoa quickly, in cases of disease, it is necessary that they should be well studied in healthy cases. This may be accomplished in the following manner:—After coitus there always remains a sufficient quantity of seminal fluid in the urethra to serve for precise and complete microscopical examination. This may be obtained by pressing the canal shortly after the act, and receiving the drop of fluid from the orifice of the glans on a plate of glass. In this drop of fluid thousands of animalcules may be seen agitating themselves like so many tadpoles in a pool of stagnant water, only that the tails of the spermatozoa are relatively longer and thinner, and that the head presents a brilliant point near its insertion. Generally the number of these animalcules prevents them from being easily examined; and it becomes necessary to spread them out by introducing a small quantity of water, and pressing firmly down the thin glass that covers them; they are found most separated on the edges of the fluid. If the water added be of the temperature of the body, their motions become free and lively, and continue so until cooling, and evaporation effect them. By avoiding these two causes of disturbance the motions of the spermatozoa may be kept up during several hours.

"However long a time may have elapsed after coitus, there are always spermatozoa in the urethra, provided they have not been washed away by the passage of urine. Although the point of the glans may be quite dry, and pressure along the whole length of the canal may not produce the least painfulness, still, on passing urine, living animalcules may be obtained from the first drop which escapes. This may be received on the glass, and is perhaps the easiest and most natural mode of obtaining spermatozoa for microscopic examination. It is evident that the same experiments may be applied in the case of nocturnal pollutions as well as in all other seminal discharges, in whatever manner they may occur. But many errors may arise from commencing with cases of disease, for it is during perfect health that the spermatozoa are most active and their development most complete, and they live longer after coitus than after any other kind of seminal discharge.

"Having thus described the means by which my microscopic observations may be verified, I proceed to show their results.

"*Spermatozoa.*—Out of thirty-three bodies which I have examined for spermatozoa, I only twice found these animalcules in the testicles. In one of these cases the patient died from the effects of a fall on the day following it; in the other acute gastro-enteritis was the cause of death. The seminal fluid was most abundant, and contained the greatest number of animalcules in the former case. The other patients died of chronic diseases after protracted sufferings. One only among them died on the second day of acute peretonitis, but he was seventy-three years of age. In thirty-one of these patients the testicles were soft, pale, and as though withered. On dissection they presented a grayish aspect, and did not furnish any liquid; the structure was almost dry, and contained a few blood-vessels; the secreting canals were easily separated from one another, and could be spread out under the microscope without breaking. They presented very brilliant granules, all of exactly the same appearance, about the size of the head of a spermatozoa, ten times smaller than the corpuscles of blood or mucus and differing from the latter by the constancy and regularity of their form. These brilliant bodies which occupied the place of the spermatozoa, are worthy of notice, because they offer considerable analogy to the appearances presented by the semen under certain circumstances.

"In order to observe what is present in the secreting canals of the testicles, it is necessary to spread out a portion of one of them under the microscope, after having examined it dry, to allow a drop of water to penetrate between the two glasses, and to follow the changes which take place, then to press down the glass so as to flatten the parieties of the canal, rupture it, and press out a portion of its contents; lastly, these must be examined again when desiccation is complete, for the spermatozoa found in the canals are then best seen.

"In the epididymis I have never found spermatozoa, except in the two cases in which they were also found in the testicles. In all the others I met with these animalcules only in the vas deferens, or seminal vesicles. There were no animalcules at all to be found in the patient that died at the age of seventy-three. It has always seemed to me that the animalcules were less numerous in proportion as the patients had suffered long; and in extreme cases I have generally found them only in the seminal vesicles. The fewer the spermatozoa, the more difficult were they of detection on account of their extreme transparency. In some cases I have only suddenly discovered them after examining for an hour or two, the liquid having previously appeared quite homogeneous. The dimensions were the same as those of the best developed animalcules, but they were pale throughout their whole extent, and more transparent than the surrounding fluid. Complete desiccation caused them often to disappear altogether; but the same phenomena could be reproduced by the absorption of a small quantity of water.

"In cases of phthisis, caries of the vertebræ, white swelling, &c., I have had great difficulty in distinguishing the animalcules, probably because those diseases do not cause death for a long time.

"I have almost always found in the seminal vesicles, especially at the bottom of any depression, a thick, grumous, brilliant matter, varying in its aspect and color, but considerably resembling thick paste, and more or less transparent; with a high power the granules of this matter appear large, irregular, more or less opaque, and without any constant shape. They are evidently the products of the internal membrane of the vesicles, for they are found with similar characters in the accessory vesicles of the hedge-hog, rat, &c.,

which never contain seminal animalcules, and do not communicate directly with the vasa deferentia, which, again, never contain any similar substance. This matter is, therefore, analogous to that secreted by the prostatic follicles, cowper's glands, &c. Its functions are the same, and for many reasons it merits special attention.

"The secretion of semen diminishes in all serious diseases, and seminal evacuations become very rare, especially towards the last. It is not, therefore, astonishing that the products of the mucus membrane predominate in such patients over those of the testicles, and that such mucus should become more consistent during its long residence in the depressions of the vesicles. Hence, the difference observable between the semen obtained from the vesicles after death, and that which is passed by a healthy person. Nevertheless, after long-continued continence, more or less large granules are often seen in the semen of a healthy person, and these are perfectly distinct from the fluid part. When the emissions are most frequent, granules of the same kind may be observed, but much smaller. These facts are important when applied to explain several symptoms of diurnal pollutions.

"I have already stated, that, on causing the patients to make water in a bath, the semen passed may be easily recognized by means of its globules, which whirl about in the middle of the cloud formed towards the close of micturation. From what we have just seen, it is evident that these globules come from the internal membrane of the seminal vesicles. They may be wanting in very severe cases when the semen has no time to acquire consistence; but their presence leaves no doubt as to the existence of diurnal pollutions, because they can only be furnished by the seminal vesicles. On the other hand, I have invariably found spermatozoa in the urine of patients who observed this phenomenon in the bath. The same remarks hold good when applied to the globules which the urine deposits in certain cases of diurnal pollutions, and which have been compared by some to grains of bran, by others to millet seed, pearl barley, &c., according to their size. These globules are perceived as soon as the urine is passed; they are roundish, very soft, and do not give any sensation when squeezed between the finger and thumb; they cannot, therefore, be confounded with urinary salts, which are deposited only when the urine has cooled, have a crystellian form, and give the sensation of a hard body to the finger. The vesicle mucus also is only deposited on cooling, and does not furnish brilliant granules.

"As to pus, its appearence is easily determined. I have found animalcules wherever these globules appeared in the urine; and hence it is that I have pointed them out as certain signs of diurnal pollutions.

"I have also noticed that in some cases the urine, when held against the light presents in the middle of a flocculent cloud multitudes of quite characteristic brilliant points. These are smaller, and consequently lighter globules than those which in other patients fall to the bottom of the vessel. They are neither observed in the mucus of the bladder nor in the prostatic fluid, which alone present clouds analagous to those of diurnal pollutions. Such brilliant points also arise from the seminal vesicles, and their presence is, therefore, an indication that the urine contains semen. This I have often verified with the microscope. I should, however, warn those who wish to repeat my experiments, that it is not in the midst of the flocculent cloud that the zoospermes are to be sought, but at the bottom of the vessel to which they soon fall, on account of their greater specific gravity.

"The results of all my observations on the dead subjects, therefore, convince me of the influence of serious and long-continued diseases on the functions of the spermatic organs. But it is not only in the morbid state that these experience great variations; remarkable differences may exist between healthy individuals, not only in the quantity of semen secreted in a given time, but also in the number, appearance and dimensions of the spermatozoa. In this respect, I have observed differences amounting to a third, and in som cases to a half. The comparison is very easily established. When the semen is kept under a thin glass, as I have described, it is not in danger of undergoing any changes, and may be always, by the addition of a drop of water, campared with a recent specimen.

"Notwithstanding the facility with which nocturnal

pollutions may be recognized, I have submitted the semen collected after them, by individuals in various conditions of health, to microscopic examination. At first, when the evacuations are still rare and the semen preserves its ordinary characteristics, the animalcules do not present any remarkable circumstances in regard to their number, dimensions, &c.; but when the disease has reached a sufficient degree of gravity to affect the rest of the system, the semen becomes more liquid, and the spermatic animalcules less developed and less lively. This number, however, does not as yet sensibly diminish; indeed, in some cases, it seems increased. As the disorder advances, the erections diminish, the semen becomes more watery, and the animalcules are often a fourth or a third less than natural, and the tail is often distinguished with difficulty, under a power of three hundred diameters. At a still later period the animalcules become fewer, and in two individuals, in the last stages of the affection, the semen no longer contained animalcules, although it retained its characteristic smell. Examined with high powers, and very proper precaution, I only found, in this semen, brilliant globules, all exactly alike, and about the same size as the head of a spermatozoa.

"The microscopic examination which I have made of semen passed during efforts at stool give analogous results. When such discharges only take place accidentally and at long intervals, the semen is thick, whitish, impregnated with a powerful smell, and abundantly furnished with well developed animalcules. I have sometimes even found a few alive after an hour or two. But when these discharges become so frequent or habitual as to constitute disease, they become less abundant and the semen loses its normal properties. The spermatozoa are generally smaller than in the healthy condition, and always less lively. I have some preparations in which they are only one-half the ordinary size, and I have never been able to find a single living animalcule a few minutes after the fluid had been expelled. When the disease has become much aggravated the spermatozoa become rare, and they are sometimes replaced by ovoid or spherical globules similar to those of which I have already spoken. In these patients in an extreme state of disease, I found nothing else, although they passed as much as a desert spoonful of semen at each stool. Such cases, however, are exceedingly rare.

"In diurnal pollutions happening during the passage of urine the following means may be employed to show the presence of spermatazoa :—The urine should first be filtered in a conical filter, when, on account of their weight, the greater number of the spermatozoa will remain on the lowest part of the paper. By taking this portion and turning it upside down in a watch-glass containing a few drops of water, the animalcules become detached from the paper by degrees, and fall to the bottom of the fluid in the glass. After twenty-four hours maceration in this position, the paper may be taken away, and the spermatozoa may be readily obtained by using a drop from the bottom of the fluid in the watch-glass for examination. This mode of proceeding is a sure one, but it requires considerable time and trouble for its performance.

"I have already stated that the urine does not always contain spermatozoa in cases of diurnal pollutions; therefore, the urine of the same individual would perhaps, require examination on many occasions, before the certainty of their presence could be established, and few medical men in active practice have time to devote to such experiments. I, for one, should have long since given up treating these patients, had I been obliged to repeat in every case such long and tiresome examinations. Ten days or a fortnight are sometimes passed without the appearance of spermatozoa in the urine, and hence all who are accustomed to microscopic researches, will admit the indefinite amount of trouble and time required. Fortunately, however, there is a more simple method by which such examinations may be conducted.

"It will be recollected that the semen always escapes with the last drops of urine, or immediately, or soon afterwards. By directing the patients, therefore, to compress the urethra immediately after micturating, and to receive the drop of fluid pressed out on a piece of glass, sufficient animalcules will be obtained from the walls of the urethra for microscopic observation. These being covered with a thin camella of glass may be either at once placed under the microscope, or may be allowed to dry, and be examined at a future time—a drop of water being previously added. This mode of examination is, therefore, easy for all practitioners who possess a good microscope, after they have accustomed themselves to the inspection of the spermatozoa in their natural state.

"The changes which I have mentioned as occurring in the semen must be borne in mind, however, and the animalcules must not be expected to appear either so large, so well defined, or so numerous, as in cases where there is no disease."

CHAPTER X.

THE QUANTITY OF SEMEN LOST; HOW IT AFFECTS THE SYSTEM, AND ITS INFLUENCE ON THE MIND.

"A too great loss of semen weakens all the solid parts; hence arise weakness, idleness, phthisis, tabes dorsalis, stupidity, affections of the senses, faintings and convulsions." Hoffman had already remarked that "those young people who practice the infamous habit of masturbation, lose gradually all the faculties of the mind, particularly the memory, and become entirely unfit for study." Lewis describes all these symptoms. We shall translate from his work only what relates to the mind:—"All the symptoms which arise from excesses with females, follow still more promptly, and in youth the abominable practice of masturbation; and it is difficult to paint them in as frightful colors as they deserve. Young persons addict themselves to this habit without knowing the enormity of the crime, and all the consequences which physically result from it. The mind is affected by all the diseases of the body, but particularly by those arising from this cause. The most dismal, melancholy, indifference and aversion to all pleasures, the impossibility to take part in conversation, the sense of their own misery, the consciousness of having brought it upon themselves, the necessity of renouncing the happiness of marriage, all affect them so much that they renounce the world—blessed if they escape suicide."

The symptoms mentioned in the foregoing are almost a perfect type seen in the cases of masturbation; but there are others more deplorable, which have but recently been discovered as arising directly from it: I mean insanity, idiocy, and a total prostration of all physical and mental power, and which affects not only the person so addicted, but the innocent offspring, and which invariably leads to the extinction of the family name.

It is but quite recent that these terrible evils have been found to be under our control. It has heretofore been considered a mysterious dispensation of Divine Providence, to be met and endured with patience and resignation. Very few eminent writers have already, though but recently, acknowledged the influence of self-abuse in producing idiocy, insanity and constitutional degeneracy, and urge the necessity of searching for the cause, in treating these evils. So convincing has this become, that it has even been recognized recently in a legislative document, which tells more wholesome truth —accompanied with more sound reasoning, I was about to say—than all the medical treatises heretofore published on the subject put together. I particularly refer to the Report on the subject of Idiocy, presented to the Massachusetts Legislature by Dr. Howe, in February, 1848, complying with a resolution of that intelligent body, directing a report on that subject. I hope the introduction of that valuable document will be the means of eliciting the like truths, by all the other Legislatures, without any farther delay, for the influence of such documents, from such sources, would accomplish more good, by preventing the unsuspecting from falling into so deplorable a vice, than all the asylums and medical treatment could ever think of doing by way of cure. I feel convinced that the time will speedily arrive when this and similar reports will be eagerly sought for, and their inestimable value universally admitted. My quotations from the report will not only corroborate all I have said in these pages, but serve the additional purpose of more fully enlightening the public.

Dr. H's able, forcible, sensible and convincing re

marks, on boldly approaching this subject, should forever silence and put to shame all affected modesty in speaking upon this subject hereafter, for too many persons, aware of its existence heretofore, have foolishly been prejudiced by false modesty from doing so:—

"There is another vice, a monster so hideous in mien, so disgusting in feature, altogether so beastly and loathsome, that, in very shame and cowardice, it hides its head by day, and, vampyre like, sucks the very life-blood from its victims by night; and it may, perhaps, commit more direct ravages upon the strength and reason of those victims than even intemperance; and that vice is

Self-Abuse.

"One would fain be spared the sickening task of dealing with this disgusting subject; but, as he who would exterminate the wild beasts that ravage his fields, must not fear to enter their dark and noisome dens, and drag them out of their lair; so he who would rid humanity of a pest, must not shrink from dragging it from its hiding-places, to perish in the light of day. If men deified him who delivered Lerna from its hydra, and canonized him who rid Ireland of its serpents, what should they do for one who would extirpate this monster vice? What is the ravage of fields, the slaughter of flocks, or even the poison of serpents, compared with that pollution of body and soul, that utter extinction of reason, and that degradation of beings made in God's image, to a condition which it would be an insult to the animals to call beastly, and which is so often the consequence of excessive indulgence in this vice?

"It cannot be that such loathsome wrecks of humanity as men and women, reduced to driveling idiocy by this cause, should be permitted to float upon the tide of life without some useful purpose; and the only one we can conceive, is that of awful beacons to make others avoid—as they would eschew moral pollution and death—the cause which leads to such ruin. This may seem to be extravagant language, but there can be no exaggeration—for there can be no adequate description even—of the horrible condition to which men and women are reduced by this practice. There are among those enumerated in this report, some who not long ago were considered young gentlemen and ladies, but who are now moping idiots, idiots of the lowest kind; lost to all reason—to all moral sense—to all shame; idiots who have but one thought, one wish, one passion—and that is, the further indulgence in the habit which has loosed the silver cord even in their early youth, which has already wasted, and, as it were, dissolved the fibrous part of their bodies, and utterly extinguished their minds.

"In such extreme cases, there is nothing left to appeal to, absolutely less than there is in dogs or horses—for they may be acted upon by fear of punishment; but these poor creatures are beyond all fear and all hope, and they cumber the earth awhile—living masses of corruption. If only such lost and helpless wretches existed, it would be a duty to cover them charitably with the veil of concealment, and hide them from the public eye, as things too hideous to be seen; but, alas! they are only the most unfortunate members of a large class. They have sunk down into the abyss towards which thousands are tending.

"The one which has shorn these poor creatures of the fairest attributes of humanity is acting upon others, in a less degree indeed, but still most injuriously—enervating the body, weakening the mind, and polluting the soul. A knowledge of the extent to which this one prevails, would astonish and shock many. It is indeed a pestilence which walketh in darkness, because, while it saps and weakens all the higher qualities of the mind, it so strengthens low cunning and deceit, that the victim goes on in his habit unsuspected, until he is arrested by some one whose practiced eye reads his sin in the very means which he takes to conceal it—or until all sense of shame is forever lost in the night of idiocy, with which his day so early closes.

"Many a child who confides everything else to a loving parent, conceals this practice in its innermost heart. The sons or daughters who dutifully, conscientiously, and religiously confess themselves to father, mother, or priest, on every other subject, never allude to this. Nay, they strive to cheat and deceive by false appearances; for, as against this darling sin—duty, conscience, and religion, are all nothing. They even think to cheat God, or cheat themselves into the belief that He who is of purer eyes than to behold iniquity, can still regard their sin with favor.

"Many a fond parent looks with wondering anxiety upon the puny frame, the feeble purpose, the fitful humors of a dear child, and, after trying all other remedies to restore him to vigor of body and vigor of mind, goes journeying about from place to place, hoping to leave the offending cause behind, while the victim hugs the disgusting serpent closely to his bosom, and conceals it carefully in his vestment.

"The evils which this sinful habit works in a direct and positive manner are not so appreciable, perhaps, as that which it effects in an indirect and negative way. For one victim which it leads down to the depths of idiocy, there are scores and hundreds whom it makes shame-faced, languid, irresolute, and inefficient, for any high purpose of life. In this way the evil to individuals and to the community is very great.

"It behooves every parent, especially those whose children (of either sex,) are obliged to board and sleep with other children, whether in boarding-schools, boarding-houses, or elsewhere, to have a constant and watchful eye over them, with a view to this pernicious and insidious habit. The symptoms of it are easily learned, and if once seen, should be immediately noticed.

"Nothing is more false than the common doctrine of delicacy and reserve in the treatment of this habit. All hints, all indirect advice, all attempts to cure it by creating diversions, will generally do nothing but increase the cunning with which it is concealed. The way is to throw aside all reserve; to charge the offence directly home; to show up its disgusting nature and hideous consequences in glowing colors; to apply the cautery seething hot, and press it into the very quick, unsparingly and unceasingly.

"Much good has been done, of late years, by the publication of cheap books upon this subject. They should be put into the hands of all youth suspected of the vice. They should be forced to attend to the subject. There should be no squeamishness about it. There need be no fear of weakening virtue by letting it look upon such hideous deformity as this vice presents. Virtue is not salt or sugar, to be softened by such exposure, but the crystal or diamond that repels all foulness from its surface. Acquaintance with such a vice as this—such acquaintance, that is, as is gained by having it held up before the eyes in all its ugliness, can only serve to make it detested and avoided.

"Were this the place to show the utter fallacy of the notion that harm is done by talking or writing to the young about this vice, it could perhaps be done by argument, certainly by the relation of pretty extensive experience. This experience has shown, that in ninety-nine cases in a hundred, the existence of the vice was known to the young, but not known in its true deformity; and that in the hundredth, the repulsive character in which it was first presented, made it certain that no further acquaintance with it would be sought."

My experience tells me, that the language of Dr. Howe is not too severe, for to speak directly to the point is what is necessary, and, as he says, "not be evasive." He never did more real service to humanity than when he presented this Report, for there cannot be more momentous truths. Ignorance is made terribly apparent, both as regards the patient and society at large, in another part of this truly valuable document. Every one will be in danger, until knowledge takes the place of such ignorance :—

"In some families that are degraded by drunkenness and vice, there is a degree of combined ignorance and depravity which disgraces humanity. It is not wonderful that feeble-minded children are born in such families; or, being born, that many of them become idiotic. Out of this class, domestics are sometimes taken by those in better circumstances, and they make their employers feel the consequences of suffering ignorance and vice to exist in the community.

"There are cases recorded in the appendix, where servant-women, who had charge of little girls, deliberately taught them the habits of self-abuse, in order that they might exhaust themselves, and go to sleep quietly. This has happened in private houses as well as in the almshouses; and such little girls have become idiotic. The mind instinctively recoils from giving credit to such atrocious guilt; nevertheless, it is there, with all its hideous consequences; and no hiding of our eyes, no wearing of rose-colored spectacles—nothing but looking at it in its naked deformity, will ever enable men to cure it.

"There is no *cordon sanitaire* for vice; we cannot put it into quarantine nor shut it up in a hospital; if we allow its existence in our neighborhood, it poisons the very air which our children breathe.

"The above remarks forcibly apply to *all* our public schools, for I have become too well acquainted, I was about to say, with the alarming extent with which it prevails, often even in the most open manner. The extent of it is amazing, for it exists both among the teachers and the students, and what can be more absurd than the partial, even shunning of the subject? By so doing, it leads not only to the continuance in some, but the production of it in the yet uninitiated.

"In some, as I have previously stated, persons commence the habit accidentally, but their numbers are very limited, compared to those who are taught it; therefore the immense importance of preventing contamination by those already addicted.

"There is one remarkable and valuable fact to be learned respecting this vice, from observation of idiots, and that is, that some of them, though they have no idea of right and wrong, no sense of shame, and no moral restraint, are nevertheless entirely free from it. They could never have been in the practice of it, else they would never have abandoned it.

"From this may be inferred, that it is a pest, generally engendered by too intimate association of persons of the same sex, that it is handed from one to another like contagion, and that those who are not exposed to the contagion are not likely to contract the dreadful habit of it. Hence we see that not only propriety and decency, but motives of prudence, require us to train up all children to habits of modesty and reserve. Children as they approach adolescence, should never be permitted to sleep together. Indeed, the rule should be—not with a view only to preventing this vice, but in view of many other considerations—that, after the infant has left its mother's arms, and become a child, it should ever after sleep in a bed by itself. The older children grow and the nearer they approach to youth, the more important does this become. Boys even should be taught to shrink sensitively from any unnecessary exposure of person, before each other; they should be trained to habits of delicacy and self-respect; and the capacity which nature has given to all for becoming truly modest and refined, should be cultivated to the utmost. Habits of self-respect, delicacy, and refinement, with regard to the person, are powerful adjuncts to moral virtues. They need not be confined to the wealthy and favored classes; they cost nothing—On the contrary, they are the seeds which may be had without price, but which ripen into fruits of enjoyment that no money can buy."

It is true, that it is almost impossible, unfortunately, to entirely prevent children from learning this vice, either by emulative practice or more direct instruction. We must necessarily then guard them against it, by a timely warning. I know of instances where parents supposed the information they had given their child, had produced the desired effect, but instead of that they have kept on with the practice, until they sunk into an untimely grave, and yet the parent was congratulating himself upon the success of his precautions.

Copland has an article on Insanity, which points out the various causes of this terrible affliction, and uses the following language, in speaking of self-abuse :—

"Many, however, of those causes which thus affect nervous energy, favor congestion of the brain and occasion disease of other vital organs, tending to disorder the functions of the brain sympathetically. Of these, the most influential are masturbation and libertinism, or sexual excesses, sensuality in all its forms, and inordinate indulgence in the use of intoxicating substances and stimulants. The baneful influence of the first of these causes is very much greater, in both sexes, than is usually supposed, and is, I believe, a growing evil, with the diffusion of luxury, of precocious knowledge, and of the vices of civilization. It is even more prevalent in the female than in the male sex, and in the former it usually occasions various disorders connected with the sexual organs—as leucorrhœa, or suppressed or profuse menstruation, both regular and irregular hysteria, catalepsy ecstacies, vertigo, various states of disordered sensibility, etc., before it gives rise to mental disorder. In both sexes epilepsy often precedes insanity from this cause; and either it or general paralysis often complicates the advanced progress of the mental disorder, when thus occasioned. Melancholia, the several grades of dementia, especially imbecility and morcomania, are the more frequent forms of derangement proceeding from a vice which not only prostrates the physical powers, but also impairs the intellects, debases the moral affections, and altogether degrades the individual in the scale of social existence, even when manifest insanity does not arise from it."

As I have asserted in a former page, the difference between the effects of masturbation and natural excess, is very great, for the former is the operation of the imagination entirely, while the latter is accompanied by the natural associations, producing a pleasurable feeling; to be sure, the latter is only partial when in excess. I very often have patients, who are unwilling slaves to th evice of masturbation to such an extent, that their hands must be fastened while at sleep, to prevent the practice. The mental tortures of fear and self-condemnation, combined with the bodily exhaustion, produce a fearful havoc. Natural excesses seldom cause insanity or idiocy, except, perhaps, in the offspring; but the solitary vice is sure to do so in both, if not prevented by judicious and timely treatment.

The Massachusetts Report says, that, "one hundred and ninety-one of the idiots examined were known to have practiced masturbation, and in nineteen of them the habit was even countenanced by the parents or nurses! One hundred and sixteen of this number were males, and seventy-five females. In four hundred and twenty who were born idiots, one hundred and two were addicted to masturbation, and in ten cases the idiocy of the children was 'manifestly attributable to self-abuse in the parents!' The ten cases known, justify the conclusion, that in reality there are many more, which proves, beyond a shadow of doubt, that many cases of idiocy in children is attributable to the sexual vice of the parents. Is not this fact almost too fearful for contemplation, and the importance of it to the community incalculable?"

There are many valuable statistics in the annual reports of the Massachusetts Lunatic Asylum. In the twelfth report, of the number of cases existing in the Institute, one hundred and thirty-nine are set down as having been caused by masturbation, which, from the language used in the report, is a decrease of the number to such an extent as to cause great satisfaction, and is attributed to the information that has been diffused on the subject, and the warnings that have reached the young through the various channels of intelligence that have been opened on this hitherto obscure subject.

The thirteenth report contains one-hundred and forty-five cases, as caused by self-abuse, with some remarks, which I add below:—

"The causes of insanity may be divided into voluntary and involuntary. Of the former, the principal are intemperance and the secret vice; other causes may be of this class, such as hazardous speculation, many religious vagaries, imprudent exposures and irregularities. None are so prominent as the two first named, and none so fully stain the character with guilt, which even the occurrence of hopeless disease can hardly wipe away. Intemperance disorders the senses, and induces apoplexy, epilepsy, and palsy. The cases from this cause are about as favorable for recovery as the majority of others, but are most sure to return if the habit of intemperance recurs.

"The secret vice produces the very worst form of insanity, because it is so difficult to avoid the continuance of the cause, and because the energies of the system are more prostrated by it than by almost any other cause. Such patients become degraded animals, so entirely abandoned to the habit that hopeless dementia and driveling idiocy generally follow. A few can be influenced to abandon the practice, and a few others can be cured in spite of it, but in almost all cases the disease will become worse, and these dreadful consequences will ensue.

"The secret vice, though doubtless a frequent cause of insanity, and of other severe and fatal diseases, far more than is generally supposed, is most operative in preventing recovery from insanity, arising from this and other causes. It is extensively and alarmingly the result of an active propensity excited by the disease, and unrestrained by reason, moral influences, or self-respect. Many cases of a favorable character progress towards recovery, till this practice is commenced, then the patient becomes listless, is inclined to lie down or sit in a bent position, walks moderately, looks feeble, and feels weak and miserable. His mind loses its energies, its scope is circumscribed, more and more, till this beastly indulgence occupies all his thoughts, and the remnant of all the physical powers is concentrated to this single effort of gross and debased animal nature. Thus, the groveling sensualist lives, often a long life, a degraded sufferer, without a manly thought or a moral feeling worthy of his nature or his destiny, and finally leaves the world without the regret of his friends, a useless, burthensome, loathsome object of abhorrence and disgust."

Under the head of "Relation of Cause to Recovery," in the two reports, there are two hundred and seventy-one males and twenty-nine females, enumerated, from the effects of masturbation; and of this number, one hundred and ninety-seven males and twenty-seven females were incurable; which shows but the small number of two of the females as having been cured.

As one of the proofs against the absurd doctrine, that by giving this subject publicity, will increase the evil, by drawing the attention of those to it, who otherwise would not have known there was such a disease, I will insert some of the truths directly relating to the occupation of those addicted to this vice, in these reports.

Those educated as physicians of course understand the sexual system, while others never study anything of the kind, for want of time or because it is not necessary. We will see by this the proportion of the educated and uneducated, as regards those made insane by masturbation.

In the thirteenth report, there were sixty-two shoemakers in the asylum, of whom twenty-four were insane from this cause; which shows that nearly one-half of this class of the population become so from self-abuse.

By referring to the medical profession, we find phy

sicians made insane from other causes, but not a single case in either report caused by masturbation. This alone is conclusive evidence, that the greater the knowledge of the subject the less will the habit be practiced, so that in the end we may almost hope for its entire extinction.

There are other facts also, in the same reports, which conclusively show, that this kind of knowledge is needed; for instance, among students the proportion of insane from masturbation is about seventy-five per cent., there being eighteen from this cause, out of twenty-five. And among merchants, the proportion is nearly fifty per cent. Of lawyers, about thirty-three per cent.; and clergymen fifty-six per cent. Intemperance is the greatest cause, yet there were many of them that were produced by masturbation in the first instance, and to drown their shame no doubt resorted to alcohol. Many, without doubt, used other stimulants as well as alcohol, as a temporary relief from the great exhaustion the practice always produces.

Self-abuse, it is well known, causes ill health, and that was the cause of some of the cases; but, notwithstanding, all of the other complaints combined do not, I am satisfied, cause as much insanity as the secret vice of self-abuse.

The influence of the occupation, as tending to this solitary vice is shown in these reports. They conclusively prove, that sedentary employments tend to cause it, while active, out-of-door occupation opposes it, for among students, printers, shoemakers, and merchants, fifty per cent. of the insanity arises from masturbation, but only twelve per cent. from intemperance. Among carpenters, blacksmiths, and others who are actively employed, thirty-five per cent. arises from intemperance, and only thirteen from masturbation. Among seamen, fifty-four per cent. of the insanity is caused by intemperance, and only eleven per cent. from solitary abuse.

Parents, therefore, after they have given their sons the proper advice, should allow them to choose their own employment, and not confine a sanguine temperament, requiring mental and muscular occupation of the most varied kind, to an office or counting-house, nor a

PLATE 23

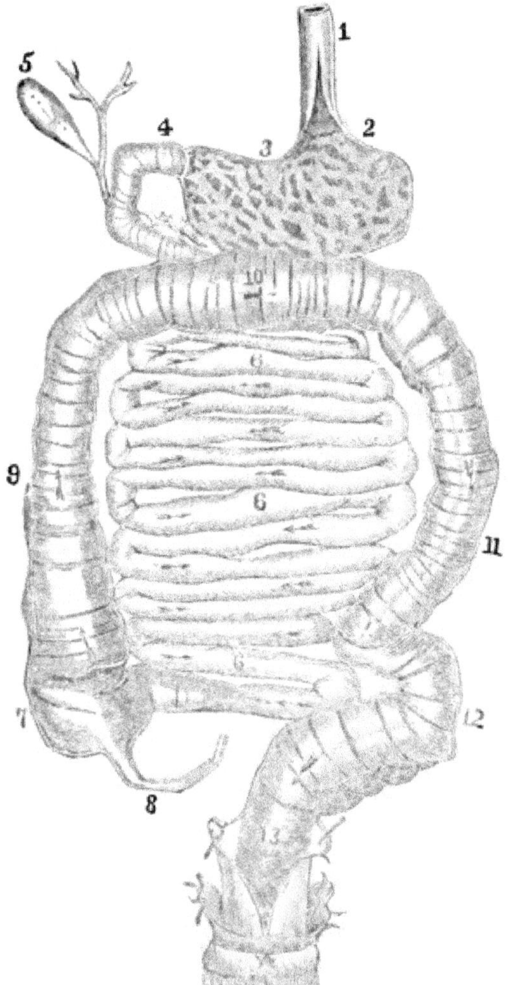

Fig. 71 represents the digestive tube from the esophagus to the anus. 1, esophagus, which is laid open at 2, to show its termination in the cardiac orifice of the stomach ; 3, interior of the stomach with its rugæ ; 4, duodenum, commencing at the pylorus ; 5, gall-bladder with the cystic duct, which last passes downwards to open into the duodenum ; 6, 6, 6, small intestine, terminating in the cæcum, 7 ; 8, appendicula vermiformis ; 9, right ascending colon ; 10, transverse arch of the colon ; 11, left descending colon ; 12, sigmoid flexure ; 13, rectum ; 14, anus.

more intellectual profession, for the monotonous inactivity, as we have stated above, is the greatest cause of sexual excesses, for the vital energy, instead of being expended in active exertion, stimulates the sexual organs to an unnatural degree, and then causes the vice, both as a gratification and relief.

The every-day cases we meet with, present these effects most frequently—weakness of the eyes, swelling and soreness of the lids, the dark or tell-tale spots under the eyes, which is a sure symptom of an exhausted system, buzzing or singing in the ears, partial deafness, weakness of the back and limbs, especially the knees, headache, dizziness, flatulence, incontinence of urine, which is very frequent, diarrhœa, but most generally constipation, and sometimes of the most obstinate form, palpitation of the heart, shortness of breath, loss of memory, and confusion of ideas, melancholy, peevishness, and irritability. Once and awhile we meet with a case where there is a partial loss of speech, tendency to stammer or stutter; and in almost every instance there is a hesitancy and indecision in these patients. Sometimes there is a difficulty of swallowing, which is a partial paralysis produced by the sympathy with the organs below, and are very often troubled with a gulping or belching of wind, a whitening of the hair, and even baldness, are often produced from these causes; palsy and epilepsy are very often the results, and so is paralysis, much oftener it is the case than is supposed. These are severe affections, but the sympathies of the generative organs are powerful, and it is not surprising, therefore, as the exhaustion of the vital power is great, when caused by excessive action. For the better illustration of the subject, however, I will here insert a few cases from the note-book of M. Lallemand, and others; but the mode of treatment I entirely dissent from, for it is very seldom, indeed, that a case presents itself, requiring, or which is adapted to it, and I have cured many a patient by my more simple treatment—which causes not the least pain—after they had submitted to numerous cauterizations, the pain of which is always severe, and sometimes causes a worse disease than the one for which it was applied.

CASE XVIII.

"M. D. of Philadelphia, of a very robust constitution, contracted the habit of masturbation while at school, when only eight years old. The first effect produced was a frequent desire to pass urine, and at twelve years of age this irritability had become so great that he was sometimes unable to retain his urine a quarter of an hour. Before entering a house he always took care to micturate several times in rapid succession; and, notwithstanding this precaution, he soon experienced renewed uneasiness. He felt as though his bladder was never entirely empty, and the smallest quantity of urine induced spasmodic contractions. The irritability of the urinary organs diminished by degrees after the period of puberty, but never ceased entirely, notwithstanding the various means which were employed on different occasions.

"At the age of sixteen, M. D. endeavored to break off his injurious habits by sexual intercourse, but he found himself completely impotent, and shame induced him to return to masturbation. He afterwards made further attempts to correct himself, but he experienced nocturnal pollutions, which often made him lose courage. At length, after many relapses, he succeeded completely, without observing any further nocturnal emissions. Still his health, instead of improving, became more and more impaired. His erections were less frequent, less prolonged, incomplete, and at length gradually ceased, together with all animal desire.

"At the age of twenty-eight, the state of his urine, its frequent discharge, and the wandering pains in the perineum and testicles, induced a fear of calculus; sounding, however, only showed a morbid sensibility of the urethra, especially towards the neck of the bladder.

"In the beginning of May, 1837, M. D. came to Montpelier, in the following condition:—Much debilitated, unsteady in his walk, easily chilled, and taking cold very quickly, wandering pains all over his body, skin dry, memory impaired, digestion difficult, extremities cold, scrotum relaxed and testicles soft—very sensitive, and often causing a dull pain, as if they were forcibly compressed; the semen (from the account he gave of the last nocturnal emisions he had experienced) clear, aqueos, and inoderous; semenal pollutions with the last drop of urine, which were clammy, and passed with difficulty, and excited a sensation of tickling in the neighborhood of the anus, which extended to the orifice of the urethra. He often had diarrhœa, but, at other times was very costive, and his stools were passed with difficulty and pain. He did not, however, often pass semen while at stool.

"I discovered, several days following, the presence of semen in M. D.'s urine, and catheterism showed an excessive irritability of the urethra, especially in the neighborhood of the prostate, which, on examination, was found slightly enlarged. Nearly a tablespoonful of blood followed the withdrawal of the catheter. The circumstances did not leave the least doubt on my mind as to the state of the mucous membranes in the vicinity of the ejaculatory ducts; and consequently, I immediately performed cauterization, from the neck of the bladder as far as the membraneous portion of the urethra.

'Twenty days afterwards, M. D. left Montpelier for Italy, and when he returned, three months afterwards, he was completely cured—no involuntary seminal emissions having afterwards appeared. His urine was transparent, and could be retained seven or eight hours, without inconvenience; its discharge took place without effort, and was not accompanied by any remarkable sensation. Lastly, the patient's impotence, which had been present nearly twelve years, had given place to a virility previously unknown to him; I need hardly state that his physical and moral energy had shared in this regeneration.

"I have often had occasion to notice the connection that exists between the spermatic and urinary organs; and I have shown that there is scarcely a cause of spermatorrhœa which does not act more or less on the bladder and kidneys.

"The cause I am now investigating affords us numerous examples of this connection—of which the case I have just related, is a remarkable instance—the irritation of the urinary organs having been developed very rapidly, having shown very marked symptoms, and having existed alone during several years. The patient was only eighteen years of age when he first became addicted to masturbation; at this early age the urinary organs alone possessed activity, and therefore they alone were able to suffer disturbance of their functions; on this account the symptoms were confined for a long time to the urinary organs. The character of the symptoms showed that they arose from a chronic state of inflammation, or from an acute irritation of the urinary organs, and this state must have extended also towards the spermatic organs. Thus, the increased sensation of the kidneys, and the extreme irritability of the bladder, would give a very clear idea of what took place in the spermatic organs at the period of puberty. As soon as the testicles began to act, they fell under the same influence as the kidneys; the seminal vesicles were in the same condition as the bladder; in other words, the semen was secreted in large quantities, and was retained a very short time in its reservoirs. Being, therefore, imperfectly formed, the usual effect on the erectile tissues produced by its presence, did not take place, and coitus was impossible at the age of sixteen.

"The occurrence of impotence at so early an age is sufficient to show that diurnal pollutions had already commenced, although the patient did not discover them for a long time afterwards. He was still, however, able to practice masturbation; and this is a circumstance which has great effect in preventing persons addicted to the vice from renouncing their fatal habits. At a later period, nocturnal pollutions, which occurred after a few days' care, shook the patient's resolution. This is a much less serious circumstance than the one just mentioned, but, at the same time, much more common. At length the patient left off his habits, and his nocturnal pollutions disappeared; yet the disorder of his health continued to increase. His prudence, exercised too late, did not arise from the strength of his will, but from the weakness of his genital organs. The disappearance of his nocturnal emissions did not arise from the remedial measures used, but from the increase of his involuntary diurnal discharges, of which he only became aware long afterwards. These common errors are the more dangerous, because medical practitioners are apt to participate in them.

"In the case of M. D. the irritability of the canal was very great, and the effect of the cauterization was correspondingly prompt and decided."

The above case is a very near type of the one from my own note-book, and which is the first case that I have presented. The age at which they commenced the habit of Onanism was the same; my own patient, however, was married soon after he arrived at age. Both cases are highly instructive and important, for, when a parent finds their child is laboring under such a difficulty in the frequent passage of urine, they should, without delay, inquire into the cause. By so doing, the accruing difficulties that will arise from the advancing disease, can be averted.

The next case also is an important one, as it shows how readily spermatorrheal symptoms may be thought to indicate almost every other disease, and, of course, the real difficulty is seldom discovered, unless the surgeon is one of practical experience in these complaints. I am indebted, (says M. Lallemand), for the following remarkable case to the kindness of Dr. Daniel of Cette:—

CASE XIX.

"On the 26th of May, 1836, I was called to F., a baker, aged twenty-two; I found him in bed, in the following condition:—Great moral prostration, carried even to a hatred of existence, prostration of strength; anaeniæ; lips pale and shriveled, remarkable palidity, eyes sunken, expression of countenance dull, great emaciation, skin hot and dry, pulse small, voice hoarse and so low that it was with difficulty a few words could be heard by approaching the ear, constant cough, scarcely permitting an instant's repose, general wandering pains, most severe in the loins and the sides of the chest, great irritability of the stomach—vomiting being excited after taking almost any kind of liquor or solid food.

"At first, I thought that I recognized in this patient the symptoms of phthisis laryngea, complicated with chronic gastritis; but the examination of his chest and abdomen did not support this opinion. The epigastric region was not painful on pressure; the respiratory murmur was heard all over the chest, and percussion emitted a healthy sound, except under the left false ribs, where it was slightly dull, and the patient felt pain.

"His debility did not permit me to practice abstraction of blood; and, indeed, the pleuropneumonia of the left side did not seem either very extensive or very acute; I therefore ordered a large blister to be applied over the affected spot, and prescribed a solution of tartar emetic and a strict diet. The pain in the side disappeared, and two days afterwards the stomach could retain milk and barley-water. Still nothing explained the patient's emaciation, his almost total loss of voice, hoarseness, and constant cough. His parents attributed these symptoms to hereditary phthisis, and mentioned that several members of the family had died of that disease. Minute and repeated examination of F's chest, however, assured me that this was not the case.

"On the other hand, the symptoms were very severe, and I could not discover any visceral lesion sufficient to account for them. In this state of uncertainty, your views on spermatorrhœa attracted my attention. I immediately questioned the patient respecting his past life, and learned that at the age of seventeen, he practiced masturbation with such fury, that he had frequently passed aqueous semen, mixed with blood. Frightened by these incidents he had corrected himself completely. But after about a fortnight's abstinence, he noticed that his urine contained a deposit of thick, whitish, flocculent matter. He never attached any importance to this, although during four years he observed it constantly, and noticed that it was more abundant after he had been much fatigued in his business. He observed also, that the last drops of urine were thick and viscid, and that a small quantity of viscid matter generally remained at the orifice of the urethra.

"His bad symptoms first commenced at this time; his erections and desires entirely disappeared, and, by the time he had attained the age of twenty-one, he was obliged to give up his employment, and shortly afterwards, his symptoms becoming aggravated, he was unable to quit his bed. I examined his urine, and found it in the condition he had described—the deposit contained in it being about an ounce in quantity. I noticed that his testicles were soft, and his scrotum flaccid. He agreed to my proposition of cauterizing the prostatic portion of the urethra, with eagerness, and I performed it on the following day. The effect of the cauterization was rapid. The second night afterwards the patient slept soundly; the third day a change was observed in the voice, and erections occurred during the night. On the fourth day the patient was able to get up and take some light food, which was well digested; his wandering pains had disappeared; and by the ninth day after the cauterization, the patient's strength had returned. Tonic regimen and the use of sea-bathing confirmed his restoration."

The next case especially exhibits that singular tendency of these diseases, if they go on unchecked, and which are sure to end in confirmed insanity. It generally commences with the hallucination, that they are about to be defrauded, that everyone are aware of their disease, consequently they are despised, and finally, that they will be assassinated, for they are certain, they think, that society generally are leagued against them, and that they are the most ill-used persons living.

CASE XX.

"At the beginning of April, 1836, M. Emile G. was sent to consult me, by Dr. Cauviere, of Marseilles. He was twenty-five years of age, and had attracted notice from the brilliancy of his intellect. At twenty-one years of age he had been admitted an advocate in a highly-flattering manner. He stooped much, and, though his bony system seemed to announce a strong constitution, his limbs were small and his muscles soft. His hair was black and thin, his skin pale, and his face without expression. His eyes were dull, and constantly cast down; his voice weak and husky, and his general appearance announced great timidity. His legs were constantly in motion.

"I learnt that M. G. had contracted the habit of masturbation at school at twelve years of age, and that, whilst studying law in Paris at the age of nineteen, he found a change in his character commencing; this I will describe in his own words—

"'At first I felt a gradually-increasing disgust of

everything, and a constant sense of ennui. From that period I only saw the dark side of life. Thoughts of suicide soon afterwards occurred to me, and this state of mind continued for twelve months, after which other ideas took the place of those respecting suicide. I considered myself a subject of ridicule, and fancied that the expression of my countenance or my manner excited an insulting gayety in the persons I met. This notion each day acquired new strength; and often, when in the street, or even when at my own house, or in a room surrounded by my relations and friends, I fancied I heard insults which were aimed at me. I think so still. At length, as my state became worse, I thought that every one insulted me, and I still think so. If any one expectorates or blows his nose, coughs, laughs, or puts his handkerchief before his face in my presence, I experience the most painful sensation. Sometimes I feel enraged; but more frequently a depression of spirits, ending in involuntary tears. I look at no one, and my eyes are never fixed on any object. Wrapped up in my own thoughts, I am indifferent to all external impressions. These signs are evidently those of imbecility. I admit that I may have had—and that I may even now have—hallucinations, but I am fully persuaded that these ideas are not without foundation. I am convinced that the expression of my countenance has something strange in it; that people read in my looks the fears which agitate and the ideas which torment me, and that they laugh at this unhappy weakness of intellect which they ought rather to pity.'

"The patient experienced a sense of heaviness and oppression in his head, and, although fatigued by slight exercise, was constantly in motion. Two years before he consulted me he began to correct himself by degrees, and for nine months he had entirely renounced the practice of masturbation; yet, notwithstanding this, his state daily grew worse. His digestion was disordered; he suffered from obstinate constipation, and his erections and venereal desires had left him for a long time. Yet he did not mention the last facts—in the written statement of his case—to me; they were minor evils; one idea alone absorbed him—the conviction that he was an object of contempt and ridicule to all who approached him; this idea was aggravated by their knowledge of his impotence and by shame for the cause which had produced it.

"This patient's urine usually contained an abundant flocculent deposit, resembling a thick decoction of barley; it decomposed very rapidly, and emitted a disagreeable smell. After every stool the point of the glans' penis was covered with a clammy, viscid matter, resembling a thick solution of gum.

"These circumstances confirmed me in the idea that involuntary seminal discharges alone opposed the patient's recovery. The frequent emission of the urine, the sensibility of the spermatic cords, of the testicles, and especially of the urethral mucous membrane and the injected state of the orifice of the urethra, made me attribute these evacuations to irritation of the spermatic organs rather than to relaxation. As, however, the patient refused to submit to cauterization, I ordered him iced-milk mixed with spa-water, cold lotions, &c., but he found himself much worse after the use of these means; all his symptoms were aggravated, and his urine became thicker, and left a glaring deposit adhering to the bottom of the vessel.

"At length, on the 23d of April, I persuaded M. G. to submit to cauterization, and I performed it immediately, chiefly on the neck of the bladder and the prostatic portion of the urethra; nothing particular occurred, except that the inflammation of the urethra, which followed the application, was not entirely removed for three weeks. This, I believe, arose, in a great measure, from the severe weather which prevailed at the time. I ordered two or three warm baths to be taken in a week, a few warm injections and demulcent drinks.

"At the expiration of a month, the patient took pleasure in going out, and occupied himself in gardening; he felt stronger, and took longer walks; he was able to employ himself longer without fatigue; he also experienced nocturnal emissions, preceded by erratic dreams and lively sensations. At this he was at first alarmed, but he gained courage when he saw that he was not injured by them.

"I had not seen him for more than a month, when one day he called on me, quite dispirited, to say that he should never get well, as he was relapsing into his former habits. I blamed him, but at the same time I explained to him that the fact was a proof of his having regained his former virility, of which he should make more proper use.

"M. G's mother came to me soon after, to speak of the propriety of marriage for her son, whom she saw exposed to various dangers. I easily persuaded her that before deciding on marriage, it would be necessary for him to be firmly assured during a considerable period of his perfect and decided recovery. M. G. had then regained his spirits, his boldness and his position in society, and eighteen months afterwards, all his functions being performed with energy, he married. Six months after his marriage I heard that his health had not for a moment been disordered.

"With this patient I received the following consultation from Dr. Esquirol:—

"'The undersigned cannot mistake a case of hypochondriasis which has lasted three years. It is evident that the nervous affection was produced by the habit of masturbation to which the patient was addicted from the age of puberty, and of which he only succeeded in breaking himself seven months since. The hypochondriasis continues very obstinately, as the cause which produced it acted for a long time, and very seriously weakened the nervous system. The undersigned attributes the little success attending medical treatment to the unfavorable weather, to the indocility of the patient, who lives in seclusion and in physical and moral torpor, and to the weakness of his mother, who allows herself to be led away by the sight of false or exaggerated sufferings. The means advised are those usually ordered in cases of hypochondriasis:—Tonics, antispasmodics, leeches to the anus, purging, change of scene, traveling, sulphuretted baths, sea-bathing, &c.'

"Dr. Esquirol sums up his opinion in conclusion, as follows:—'I must repeat what I have said above—weakened evacuation is the cause of the disease, and everything which can strengthen the nervous system, will be useful. It was clear that masturbation had been the first cause of the physical and moral derangement, called hypochondriasis; but the patient had renounced this vice during nine months, and his state became worse daily, instead of improving. It was evident, therefore, that some other cause acted in keeping up the disorder; and it was just as evident that this cause was involuntary diurnal discharges.

"It is not necessary for me to show that masturbation can, acting alone, induce involuntary discharges, or that the cure was due to cauterization only, although its effects were not manifest for a month after the application of the caustic; but I must insist on the pathological condition of the genital organs exciting these involuntary evacuations, since they have been too frequently ascribed to a state of debility or relaxation of the tissues. The tonics ordered by Esquirol had produced no benefit.

"'I have described the symptoms which led me to suspect acute irritation of the prostatic portion of the urethra, and I have shown the injurious effects of cold lotions, iced-milk, spa-water, &c. It was then, not by causing contraction of the orifices of the ejaculatory ducts, that the cauterization produced its beneficial effects, but by dispersing the chronic engorgement of the mucous membrane. The advantage derived from warm baths during convalescence corroborates this opinion.'

"In M. G's case a predominating symptom attracted the attention of the practitioners; hence they looked on the disease as being hypochondriasis, monomania, or hallucination, continuing after the separation of its exciting cause, and becoming consequently, an idiopathic affection. I have, however, shown that all the functions had been altered more or less; I should add, that the digestion was the last to be re-established perfectly. Such mistakes are very common and very serious, and I cannot too strongly impress their importance on the attention of the profession.

"Esquirol justly stated, the hypochondriasis took its origin from masturbation; that the nervous system was weak and excited. But he mistook the cause which kept up this condition of the brain. When masturbation has not induced involuntary seminal emissions, recovery soon follows, on leaving off the habit which has destroyed the health. Within a week the patients begin to experience a notable improvement, and in a very short time they are hardly recognizable, whatever may have been the degree of weakness to which they were reduced.

"But when Dr. Esquirol wrote his opinion, seven months had elapsed, during which M. G's conduct had been irreproachable, and when I saw him two months after, his state was even worse, although he had never resumed his former habits. The symptoms were, however, kept up by involuntary diurnal discharges.

"The effects of cauterization were very conclusive, and so soon as its curative action was felt, the patient of his own accord, took various kinds of exercise, and sought out the different amusements which had been in vain ordered for him previously; he entered into society, and did, without being pressed, all that he had before refused to do; his ideas, and his necessities altered in proportion as his functions were re-established.

"It is in vain that we say to the so called hypochondriæ—Amuse yourself, employ your mind, go into society, seek agreeable conversation; so long as we have not removed the cause of his disorder, he is unable to profit by our counsels. How can we expect that when a man is fatigued by the least exercise, he shall occupy himself with walking or gardening? How can we desire him to go into society, when the simple presence of a woman intimidates him, and recalls all his former misfortunes? How can we expect him to enjoy conversation when he loses its thread every moment? When his memory leaves him, and he feels his nullity? We persuade him to seek amusements and pleasures, but are they such to him? Is not the happiness of others, his greatest punishment? Because he is unable to follow our advice, we accuse him of unwillingness, and we wish to compel him. Let us first remove the cause of our patient's disease, and we shall soon see that his character and conduct will change, and that he will return to his natural tastes and habits.

"It is not long, in such cases, before we are embarrassed by questions about the propriety of marriage being put to us. This is a matter which is serious in all its aspects, and on which the least scrupulous should not pronounce, without having had sufficient assurance of their patient's health is now not the only one, nor is even his future happiness alone implicated; the fate of the innocent being who is about to be associated with him is the matter of chief importance, and justice to her demands that we do not counsel matrimony until sufficiently long proof has been given that our patient's re-establishment is permanent."

The above case will satisfy every one who are afflicted, or who have been subject previously to the habit, or to the emissions, and may be, who have married, without having first been cured, but who think they are well enough; yet they do not feel exactly as they should; or they may have been under treatment, and supposed themselves conval-cent; but yet the testicle on the left side is weaker d, and hangs lower than the other, or than it shou'd; the veins, no doubt, are enlarged, or when yo take the testicles in your fingers there is a feeling like worms collected together. Both testicles may involved, and sometimes they swell and feel tender specially if the person has been indulging in spirit us liquors, or has a severe cold; the desire for con on finally grows less; the person becomes weak, will soon go to an early grave, probably an idiot. All through these different stages there is either a visible or invisible loss of semen, which must be stopped, and as I have just said, should satisfy every one there is no time to be lost, or they may be past recovery, for, so long as the spermatorrhœa continues, no cure can be expected, but, on the contrary, the patient must go on from bad to worse, notwithstanding all that can be done for him. In addition to what I have said as to the origin of this vice, I will insert what M Lallemand says on the same subject:—

CHAPTER XI.

"CAUSES OF ABUSE

"These may be divided into two classes:—First, causes inherent in man, or those acting from within; these may be considered as predisposing causes;

secondly, external causes, or those arising from accidental circumstances, and these may be considered as exciting causes.

"*Internal or predisposing causes.*—Of the first class of causes, the most important is undoubtedly due to the human organization. In the lower animals the male and female live together, as if there were no difference of sex, except during the short rutting season. This period passed, perfect calm is restored. In the human species, the secretion of semen constantly goes on, from the time of maturity until extreme old age; the secretion may, indeed, be increased or diminished by excitement or repose of the organs; but, during this period it is never entirely suspended as long as the secreting tissues are healthy. Still, this universal and important fact has been much neglected; its application is evident.

"The form of the superior extremities in the human race also possesses a considerable influence in predisposing to abuse. Many animals are always fit for fecundation, spermatozoa being found in them at all seasons. They are, however, unable to excite seminal impressions without the aid of the female. Other animals, again, which, during the rutting season, show an almost incredible amount of erotic fury, are still unable, by their own actions, to cause spermatic discharges; their form alone prevents this, for they often attempt it, and a few even succeed. It is well known with what fury apes are addicted to masturbation—the ape being, of all the lower animals, the nearest to man in form. To this original disposition, more perfect in man than in any other animal, must be added the influence of pathological causes.

"I have already spoken of the irritation caused by ascarides in the rectum, of the erections they excite, and of the abuses induced by them. We shall see, by and by, that herpetic eruptions on the penis and prepuce may produce the same effects, and I shall show also, that an accumulation of sebaceous matter, between the prepuce and glans, may have a similar influence. I must also mention irritation of the cerebellum as inducing serious abuse, of which I shall give cases in their proper place.

"There is even some connection between the organs

of generation and distant diseases; for Dr. Desportes has mentioned a kind of angina which is frequently preceded by a considerable increase in the venereal desires, and, consequently, by a disposition to all kinds of abuses. Pulmonary phthisis also is often attended by considerable venereal excitement. It may as well then be at once admitted, that causes predisposing to masturbation exist in the human organization itself.

"*External* or *exciting causes.*—Of these I shall lay particular stress on such as act before puberty, because they have hitherto attracted very little attention. The most anxious parents believe that there is no occasion to watch over the actions of their children with regard to their genital organs, previously to the epoch of puberty; and few, even of our own profession, are led to suspect bad habits before that period. This is a fatal error, against which it is necessary to be on our guard; numerous causes may give rise to abuses at a much earlier period—infancy being hardly exempt from them

CASE XXI.

"I saw one unfortunate child, which, while still at the breast, nearly fell a victim to the stupidity of its nurse. She had remarked, that handling the genital organs appeased its cries and induced sleep more easily than any other means, and she repeated these manouvres without noticing that the sleep was preceded by spasmodic movements. These increased and took on a convulsive character, and the child was losing flesh rapidly and becoming daily more irritable, when I was consulted. At first I attributed the disorder to worms, teething, &c., but my attention being attracted by certain signs, I examined the genital organs, and found the penis erect. I was soon told all, for the nurse had no idea she was doing wrong. It was necessary to dismiss her, for her presence alone sufficed to recall to the child's memory sensations which had already become a habit. Time and strict watching were required before these early impressions were entirely effaced.

"Dr. Deslandes relates two similar cases, and Promar Halle, in his lectures on hygiene, used to mention such; Chaussier too told me of several that came under his notice; and both these observers believed such cases to be less rare than they are usually considered. These manouvres quiet the children very readily, and nurses always endeavor to obtain quiet at any sacrifice; they have no idea of the consequences of their conduct. At a later period, children are exposed to the same dangers, on the part of the servants having them in charge; and in these cases it is not of ignorance that the attendants are to be accused. Many patients have consulted me who owed their disorders to this cause; and in Case 61, I have shown the influence which such early habits exerts on after-life.

"In some children there is a kind of precocity of sexual instinct, which leads to very serious results. In these it often happens that the sexual instinct arises long after puberty; such children manifest an instinctive attraction towards the female sex, which they show by constantly spying after their nurses, chambermaids, &c. These freaks of children are usually laughed at, but if they were regarded with more attention, it would become evident that the sexual impulse has been already awakened.

"Rousseau, in his confessions, has well described the influence which early sexual impulse exercised on his whole life, and I have received numerous confidences of the same nature, which, however, it would be of no service to relate here. One case, however, is so remarkable, that an abstract of it may be instructive:—

CASE XXII.

"M. D., the son of a distinguished physician, between five and six years of age, was one day in summer, in the room of a dressmaker who lived in his family; this girl thinking that she might safely put herself at her ease before such a child, threw herself on her bed, almost without clothing. The little D. had followed all her motions, and regarded her figure with a greedy eye. He approached her on the bed, as if to sleep, but he soon became so bold in his behavior, that, after having laughed at him for some time, the girl was obliged to put him out of the room. This girl's simple imprudence produced such an impression on the child, that, wh he consulted me, forty years afterwards, he had not forgotten a single circumstance connected with it. The continual occupation of his mind by lascivious ideas, did not produce any immediate effect, but, about the age of eight, the most insignificant occurrence served to turn his recollections to his destruction. Having mounted one day on one of the movable frames which are used for buckling coats, he slid down the stem which supports the transverse bar, and the friction occasioned caused him to experience an agreeable sensation in his genital organs. He hastened to remount and to slide down in the same manner, until the repetition of these frictions produced effects which he had been far from anticipating. This discovery, added to the ideas constantly before him, gave rise to the most extraordinary abuses, and, after a time, to excessive masturbation.

"I need not mention all the miseries which followed this fatal passion; it will be sufficient for me to relate the means to which he had recourse for its correction. He slept on a very hard bed without a shirt, in order to avoid friction, and covered by a single coverlet sustained by a cradle; his arms were raised, and crossed above his head; a servant remained by his side during the night, with orders to awake him if he changed his position. When he got up, he put on next his skin a shirt of mail weighing twenty-two pounds, resembling those worn by the knights of old, except that it had no sleeves, and that it was attached, at its lower extremity, to a silver basin, fitted to receive the genital organs, and provided with openings for the thighs. This shirt of mail was in front, in order to be easily put on and taken of; and when on, it was laced up with a steel chain, a padlock being attached to the end, the key of which was kept by the servant, who had orders not to give it up on any pretence whatever. Guarded by the silver basin, the genital organs were completely removed from the touch, a little opening only being left for the discharge of the urine. As a still greater precaution, the patient had caused four sharp points to be fixed in front of this case, in order to directly oppose any erection. This apparatus he continued to wear for nine or ten years, although it frequently caused inflammation of the testicles and spermatic cords, by its pressure. Notwithstanding all these precautions, the patient's moral and physical condition were deplorable, which led me to suspect the presence of diurnal pollutions.

"I should observe, that in all cases of which I have just spoken, the children were five or six years of age—at most eight—that they did not show signs of puberty for several years afterwards, and that they were not exposed to the influence of bad example. Their sexual ideas were, therefore, spontaneously developed several years before the development of the genital organs. The same precocity is often observed in children of the other sex. Of this I shall treat more fully hereafter; at present

CASE XXIII.

I shall merely call attention to the case related by Parent Du Chatelet, of a little girl, who, after the age of four years, gave herself up to the most unbridled abuses.

"From these facts, an important scientific conclusion may be adduced, viz: that in many children the genial instinct shows itself with much energy, many years before the age of puberty.

"A no less important practical precaution presents itself, viz: that the age of puberty should not be waited for, in order to surround children with prudent circumspection and to prevent their curiosity from being gratified.

"Many parents are remarkably careless on the latter point; they permit children of both sexes to play together, promiscuously for hours, without any surveillance, provided that they are removed from all danger of accident, and that their noise is not annoying. The confidence of many parents, also, in the ignorance of their children, makes them careless of the marks of familiarity which are given to each other in their presence; children's sleep is not always so real or so sound as it seems.

"It is sufficient to point out these facts—every person can deduce the conclusions; and now I hasten to consider a question, the gravity of which has been allowed by all who have written respecting masturbation —I mean, the influence of example in educational establishments.

"If I may judge from my own observation out of ten persons whose health has been deranged immediately or recently from the effects of masturbation, nine first contracted the habit at school. All that I have read on the subject has led me to conclude that this proportion is not exaggerated. A child brought up in the bosom of his family is, it is true, surrounded by many causes sufficient to arouse his curiosity and excite his imagination; but such causes act accidentally, and in an isolated manner—they only produce a serious effect on a few ardent imaginations; a thousand circumstances may remove the attention from them. At school it is admitted that such causes do not exist; but there are others less numerous and less varied, but which operate in a much more active and continuous manner; the effect of these are direct and almost inevitable. The child finds on his first arrival a focus of contagion which soon spreads itself around him; the vice is established endemically, and is transmitted from the old pupils to those newly arriving. If a few privileged individuals escape being initiated, they are only such as do not experience any gratification. But their time will come at a later period—when the passions make themselves felt—the same circumstances will be presented to the mind under a less disgusting aspect.

"I shall not enter into details on this subject; but from all that has come to my knowledge from various and direct sources of information, I do not hesitate to affirm that nowhere are obscene books circulated more freely and boldly than in educational establishments; that the origin of the vice is not solely in the scholars, but also in the ushers and servants; that the abuses are not always confined to masturbation, and that they are not always propagated by example or persuasion, but are sometimes enforced by threats and violence. Let it not be thought that I am now speaking of rare and exceptional cases, or that I exaggerate—I possess multiplied and convincing proofs of my assertions; I would not either that I should be misunderstood. I am far from denying the advantages of education in a public school, and I am ready to admit that the competition among a number of children produces emulation, forms the future character, early shows each his own value, and lays the foundation of friendships which endure through life."

A too sedentary life is injurious at all ages, especially in childhood, when there exists such constant desire for exercise and change. Gymnastics, therefore, should on this account alone occupy an important position in the system of education; but they must be viewed under a much more serious aspect. Nothing can prevent the genital organs, at the time of their development, from reacting on the economy and giving rise to new sensations and ideas. It is impossible to prevent the attention from being attracted by the impressions caused by these organs; impossible to restrain the imagination and to prevent it from frequently dwelling on such impressions. The slightest circumstance may in such a case, lead to a fatal discovery, even if the information be not transmitted directly and enforced by example. How are such discoveries to be prevented, or rather how are their results to be guarded against? Study gives us no aid here; indeed, the continued sitting necessarily heats the organs, already too excited. The eyes may be fixed on the book, the ears may appear to listen to the master, but who can guard against the wandering of the imagination? At night it is still worse; no surveilance can prevent this. There exists only one means capable of counteracting it, and that is, muscular exercise, carried so far as to induce fatigue. This alone is able to deaden the susceptibility of the newly acting organs which excite the economy; exercise alone, by requiring matter for the repair of the muscular waste it causes, withdraws a stimulant from the genital organs, and induces sound and refreshing sleep.

Dr. McDonald says, M. Lallemand speaks of the colleges and private schools in France, but that he regrets to say that his statements apply with nearly their whole force to the schools of England. Vice is common in them; neglect of physical education, and the contracted nature of the studies to which pupils are confined in our classical seminaries—the understanding being unappealed to, and the reasoning faculties unexercised—the natural sciences neglected, and the whole of the pupil' life until the age of seventeen, employed in the study of the dead languages—are matters of vital importance, to which society has only recently begun to direct its attention. He says also, that M. Lallemand enters very fully on the subject of education as conducted in France, and well exposed the errors of the system. Most of his remarks apply to our own educational system, yet as the subject is not strictly medical, and as, moreover, M. Lallemand has treated it at considerable length, I think it best to refer those of my readers who may wish information on it, to the original work.

These remarks will apply to the schools everywhere in all other countries, and especially can I say so in respect to the schools on this continent.

CHAPTER XII.

VARIETIES OF ABUSE.

"I think it will be useful for me to give a few details respecting the different kinds of abuse which have come under my notice, and of which I have seen the hurtful influence on the genital organs. I shall omit all such remarks as have not a strictly practical bearing.

"We have already seen the dangers to which compression of the urethra, to prevent the discharge of semen during ejaculation may give rise, (Case 35), of Lallemand. In the case I have related, it seems likely that a rupture took place in the mucous membrane, because the patient felt, at the instant, an acute pain, and the following day a discharge commenced, which continued until the application of the nitrate of silver. Soon after the commencement of the discharge involuntary seminal emissions occurred, attended with serious symptoms. It was immediately behind the glands that this patient compressed the urethra, and it is quite conceivable that the sudden and violent distention of the canal might cause a tear in the mucous membrane; but this is not always the case. One of my patients writes as follows:—

CASE XXIV.

"'At the age of fourteen I practiced masturbation three or four times a week, and sometimes frequently during the day. In order to prevent the discharge of semen, I compressed the root of the penis firmly. Nothing escaped at this time, but I soon observed that the semen was discharged with my urine the first time I passed it. I followed this practice for about two years.'

"Diurnal pollutions soon appeared, and grew more and more serious. The remainder of the case presents nothing which is not met with in all cases of spermatorrhœa. What I wish to call attention to here is, that the compression was made close to the orifice of the ejaculatory ducts, and that the patient thought at first that his manœuvres were not followed by any loss of semen, although he at length discovered the contrary. Fournier and Begin report a similar case:—

CASE XXV.

"It was that of a young man, who, at the moment of ejaculation compressed the most remote parts of the urethra, so that not a single drop of semen could escape; yet the result was the same as in ordinary cases. Notwithstanding his precautions, his strength diminished, and his disorder made just as rapid progress as if the seminal emission had been perfect.

"The following is even a more remarkable case. I shall allow the patient to speak for himself:—

CASE XXVI.

"'I am thirty-two years of age, and I have had nocturnal emissions from the age of fourteen; I have also suffered from discharges while at stool, for ten years. The cause of these pollutions cannot be referred to masturbation, for I have not practiced it twenty times during my whole life. The pollutions are rather owing to reading obscene books, for they commenced soon after. At first, ejaculation was preceded by dreams, and accompanied by active erections and acute sensations, the semen being ejaculated with force. I tried various means to prevent these discharges. I have slept, during whole nights, with my penis dipped in cold water, or compressed between two pieces of wood formed for the purpose. I have tried to keep myself awake in order to prevent an emission, because, when I succeeded, the following day I felt stronger; but, after two or three nights, sleep always overpowered me. I often awoke, however, in sufficient time to prevent the catastrophe of my dreams, but frequently it was too late; on such occasions, to delay the discharge or to render it less copious, I compressed the base of the penis firmly; but it seems that these compressions greatly injured the parts, without preventing or diminishing the discharge, which took place inwardly, as I have often been convinced by inspecting my urine. From that period the pollutions have no longer been preceded by dreams, and the sensations have left me, so that I am not now aroused from sleep. My erections diminished, and have even latterly ceased entirely. For three years erections have rarely accompanied the emissions; when they do occur, I am always less fatigued. There is one thing which I have not been able to understand, and which will, without doubt, appear absurd to you; it is, that I experience pollutions without erection, sensation, or the escape of semen by the urethra. I believe that the discharge passes in a retrograde direction and becomes mixed with urine, because the next morning I find little globules, a cloud and filaments in that fluid, just as formerly, when I prevented ejaculation by compressing the root of the penis; whilst my urine contains nothing during the day or the next morning, when I have not experienced these pollutions. On waking I am perfectly aware of what has occurred, by the sweat that covers my face, the fatigue I feel in all my limbs, the headache and dizziness that affect me, the dark circles that surround my eyes, &c. I have tried cold and iced applications with slight benefit. For some time the pollutions were rarer, and were accompanied with erection and sensation; but soon they became as before, and emission did not take place outwardly. These internal pollutions have always been the most weakening. Whenever I succeed in passing the night without sleep, my urine is transparent in the morning, and I feel strong. After several nights without sleep, I generally have an energetic emission, which fatigues me little; but soon those without erection and without external discharge return, and then I always feel worn out on waking.'

"This patient's medical attendant would not believe in the possibility of pollution without the external discharge; but it seems clear that the patient really had internal emissions without perceptible discharge; that is to say, that the semen passed into the bladder, and was discharged with the urine, as had occurred before, when ejaculation was prevented by pressure on the pereneum. This compression was made in front of the ejaculatory ducts, and was very often repeated. It seems therefore likely that it was the frequent repetition of these manœuvres that at length caused the spontaneous passage of the semen into the bladder. But this is a question to which I shall have occasion to return.

"Yet all these manœuvres scarcely differ from the various means recommended by some surgeons for preventing nocturnal pollutions;. and we may then perceive how little confidence is to be placed in the instruments invented for that purpose, and the inconveniences to which they may give rise. It seems likely that the dangers would be nearly the same, in whatever part of the penis the compression is made; except if there be sufficient space in the urethra, between the point compressed and the ejaculatory ducts, to contain all the semen, it would be discharged directly the compression is removed. When, on the other hand, the compression is made immediately in front of the orifice of the ejaculatory ducts, the semen flows back, at least in a great measure, so as to induce the patient to believe that the discharge had been stopped, or, at all events, in a great measure diminished, and to induce a degree of security which leads to further abuses.

"But to return to the description of the abuses which have been admitted to me by so many other patients:—

CASE XXVII.

"One of these informed me, that about the period of puberty, while hanging one day by his arm, he experienced an energetic erection, accompanied with pleasure, and that by his efforts to raise his body, he caused an abundant seminal emission. This was the first. The next day he repeated the same motions, and noticed the same phenomena, and from that time he knew no other pleasure. From the principles which had been early instilled into him, he would have thought himself degraded by connection with a female, or by the least manual contact with his genital organs; but his conscience was quiet with regard to these practices, because they had not been forbidden him. He continued, therefore, to hang by the hands, from the furniture, doors, &c., without being suspected by any one, and fell by degrees into a state of debility and wasting, equal to those caused by the most unbridled masturbation. After a time, from weakness the patient lost the power of hanging, and his voluntary emissions ceased; but they were soon replaced by nocturnal emissions, which were very difficult of cure.

CASE XXVIII.

"The following are a few passages from a letter I have recently received:—'Being of an ardent temperament, I abused myself from the age of eight years, by practicing masturbation, or rather, by still more hurtful manœuvres. By compressing the penis between my legs, or against the seat on which I was sitting, I produced excitement, which was commonly followed by the discharge of a few drops of viscid and transparent fluid. This practice I repeated several times a day, up to the age of sixteen, when I ceased entirely, having been frightened by the discharge of nearly pure blood, which occurred several times. From this time I only sought natural enjoyments, but I found it impossible to obtain a complete erection. This state was attributed to weakness, and was combated by tonics, stimulents, and even irritants of all kinds, which have done me much injury. I used also cold bathing and cold lotions

CASE XXIX.

"I have seen an officer of high rank who had fallen nto the same condition, from the practice of similar manœuvres. He experienced the first sensation against the leg of the table, at the early age of ten years, and continued for several years to employ the same means.

CASE XXX.

"I have already related the case of another child, who allowed himself to slide down a wooden pole, and the deplorable influence which this circumstance exercised on the remainder of his life.

"In a few of my patients, horse-exercise caused the first seminal emissions. I shall relate, by and by, the case of one of those who knew scarcely any other pleasure, and who became quite impotent at the age when virility generally is the greatest. The extreme susceptibility which the genital organs manifest at the period of puberty, should prevent horse-exercise from being commenced about this time, as is usually done. It should be begun a few years earlier, or a few years later.

"I have already spoken of the danger of allowing children to sleep on the abdomen, (see Case 33, Lallemand.) I should add, that many of my patients thus contracted habits which ruined their health. Independently of the inconveniences to respiration, digestion &c., which arise in this position, erections are favored. The least friction awakens new sensations, and, once on the track, progress is soon made. Sometimes recollections have caused the choice of this position; of this I have related a remarkable example, (see Case 34, Lallemand); at other times, scruples early instilled by a sage foresight, but which the violence of the impulse has at length succeeded in eluding, have induced it.

CASE XXXI.

"Thus, I have been told, respecting one of my patients, that he would suffer death rather than defile himself by touching the genital organs; yet, for five or six years, he seldom passed a night without working his own destruction while lying on his abdomen. It is not necessary for me to enter into a description of the other means by which patients have sought to satisfy their genital impulses, without transgressing the religious and moral principles which had been taught them from infancy. Suffice it to say, that if they have succeeded in satisfying their consciences, they have not succeeded in preserving their health.

"But to abstain from all direct action on the genital organs, is not always sufficient to preserve the patient from serious disorders. A purely nervous excitement, awakened by other senses, or directly produced by erotic, may bring the same results as the worst abuses, if prolonged or repeated erections are caused by it. The following are a few such examples:—

CASE XXXII.

"A student aged twenty-two, born in Switzerland, of sanguine temperament, and great muscular power, fell into the most complete state of impotence, after having been for some time exposed to ungratified excitement. He had never practiced any solitary vice; but violent and prolonged erections came on, and were produced during the day by the influence of the memory. These erections caused abundant and frequent nocturnal emissions. Absence put an end to the excitement. The nocturnal pollutions diminished by degrees, and at length ceased entirely. Yet this patient fell into the same state of impotence as if he had committed the greatest excesses in masturbation, and at the same time preserved the appearance of health and strength. The cause of his impotence was evident on examining his urine, and causing him to watch for diurnal pollutions while at stool, but the cure of these pollutions was only perfect after two years' treatment.

CASE XXXIII.

"I have seen another case of the same kind, in a young man who passed from a state of habitual priapism to one of absolute impotence, without any other cause than violent excitement of the genital organs by an ardent attachment; he had never given way to excess of any kind.

CASE XXXIV.

"I also had under my care an English officer, who left Calcutta in perfect health, and arrived in London completely impotent, after having suffered during two months from almost constant excitement, caused by the presence of a female on board ship. This state, so opposed to that which had preceded it, continued for two years—the whole of this time not being marked by the least sign of virility. It is scarcely necessary to add that this state was produced by diurnal pollutions.

"I related a case a few pages back, in which nocturnal pollutions were caused by reading an obscene book; and I have seen a multitude of cases of this nature. From these I conclude, that in certain very excitable individuals, reading such works, the sight of voluptuous images, lascivious conversation, in a word, all things that can excite or keep up irritation in the spermatic organs, are capable of producing the same effects as actual abuse, even when the will is sufficiently powerful to prevent the thoughts from leading to the acts. On the other hand, an abundant secretion of semen with importunate erections, irritation of the urethra and prostate, always results under such circumstances, and those favor the occurrence of nocturnal and diurnal pollutions, as serious and perhaps more difficult of cure than those produced by masturbation, because it is impossible to act directly on the memory or imagination.

"It is not sufficient then to prevent all material action on the genital organs; it is necessary also to prevent all erotic excitement of the senses and all concentration of the ideas on lascivious objects. Fortune's favors are so distributed, that numbers live in absolute indolence without being blamed by the world, because they demand nothing from any one. This inaction produces results, the only remedy for which that I am aware of, is daily fatigue of the body by various kinds of exercise."

CHAPTER XIII.

EFFECTS OF ABUSES.

The effects produced by the different kinds of abuse of which I have been treating, vary according to the age of the patient, his idiosyncrasy, and the different organs chiefly affected. I have laid particular stress on the causes which may lead to bad habits some time before puberty. I must now consider their effects during this period.

"The symptoms arising from masturbation in the child, (says Lallemand), have been always hitherto confounded with those produced in the adult; they present certain distinctive characters, however, which require our consideration. However young they may be, children lose flesh and become pale, irritable, morose and passionate; their sleep is short, disturbed and broken. They fall into a state of marasmus, and at length die, if not prevented from pursuing their course. Examples of such a termination are so well known that I forbear to quote them.

"Analogous symptoms are shown in the adult—follow nearly the same course—and may lead to the same termination; but in infancy, more or less severe nervous symptoms are superadded, which are not found in those who commenced the practice after puberty, or which at least are not in the latter case manifested to the same extent; such are spasms and partial or general convulsions, eclampsia, epilepsy and paralysis, accompanied with contraction of the limbs. These phenomena were present in all the children whose cases I have noticed, and numerous similar facts have been published by different authors. Contractions of the limbs have been well investigated by Dr. Guerston, and he notices that they especially affect such children as are lank, unhealthy looking, nervous and worn out by bad habits. The following case is sufficiently remarkable:—

CASE XXXV.

"In 1824, a woman brought her son, eight years old, to the hospital St. Eloi; he had lost the use of his lower extremities for some months. The limbs were fixed, drawn together, and all the muscles contracted. The child was extremely thin, and his intellect was much disturbed. Masturbation, the cause of all these disorders, had only been discovered by his mother a few weeks before she placed him under my care, but she had used every means she could devise to prevent it without effect. After two or three trials I found that it was of no use trusting to the straight waistcoats and other means usually employed, and accordingly I determined to pass a gum-elastic catheter into the bladder, and to fix it so that the patient should be unable to withdraw it. The presence of the foreign body excited inflammation of the urethra, as I expected; when this occurred I withdrew the instrument, but replaced it as soon as the inflammation had subsided. I kept up in this manner a constant state of inflammation for a fortnight, which rendered the parts so painful that the child was unable to touch them. This treatment produced more decisive success than I had ventured to hope. Within eight days the lower extremities had regained sufficient strength and mobility to allow the child to get up; and in another fortnight he was able to run about the yards I then sent him away, threatening him with a return of the same treatment if he relapsed. The pain caused by the catheter seemed to have removed all the other impressions, for his health continued good, and growth followed its ordinary course.

"I have since employed the same means in many cases with just as much success, and I think it more sure than any other, because it is impossible to rely on the patient's will or on the assiduity of those who are appointed to watch over him. In children too it leaves an impression on the memory which is often sufficient to destroy the empire of habit and to prevent a return to the former manœuvres.

"But to resume the consideration of the symptoms observed in children: In childhood, seminal emissions are never experienced, but, nevertheless, the patients fall into a state of marasmus, to which some even succumb. These effects, like those observed under the same circumstances in the female, have induced some authors to leave out of their consideration the seminal discharges which are produced by the same acts at a later period. They have attributed the debility, which follows all abundant discharges of semen, to the nervous excitement and convulsive motions which usually accompany the discharge.

"The accidents observed before puberty are evidently only due to the effects on the nervous system, and the same sensation accompany voluntary emissions after puberty; it is natural to suppose that the nervous system plays as active a part then as in childhood. I willingly admit the importance of this nervous exhaustion in whatever manner it may be supposed to operate; and, supposing even that its action on the economy is just as important as during childhood, (which is not the case, as I shall presently show). This is no reason why the actual discharges should not be taken into account, seeing that they greatly modify the character and consequences of the nervous disturbance.

"I have already noticed that the symptoms produced by abuses during childhood, present a spasmodic character; this character, without doubt, is derived from the predominance of the nervous system at that time, rendering children so alive to external impressions. This excessive sensibility also explains the great disorder of the economy which children suffer from such manœuvres.

CASE XXXVI.

"Deslandes relates a case, showing that any action of the same kind may produce the same effects at this early age. He says, 'An observer worthy of credit, Dr. Nurambeau, has communicated to me the case of a child who procured himself similar sensations by drawing out the navel. His health became much disordered from the effects of this strange habit, which had such a power over him that coercive measures were required for its correction. It is worthy of remark that this patient showed neither erection nor any other phenomenon of the generative organs which at all referred to sexual intercourse. The organs of generation, therefore, had no influence in producing the sensations experienced by this child; but the repeated titilation of a very sensitive part produced the same disorder as masturbation.'

"It was proved in the debates on a recent criminal trial, that death may be caused by prolonged tickling of the sole of the foot. Nervous disorders, arising from such proceedings, may then be carried so far as to cause death, and from this may be imagined the effects of the multiplied convulsive shocks which irritable children produce, by acting on the most sensitive organs in the economy. Every excessive loss of semen also, even when unaccompanied by sensation, is followed by debility, and this may be carried so far as to cause death. I have related several such cases in the beginning of this work.

"There exists, then, two distinct causes—nervous disturbance and debilitating discharges, and both these act at once when seminal emissions are produced by the influence of the will. It is not to be wondered at that both these causes should produce nearly the same symptoms, because they both weaken the economy. The action of the first on the nervous system is direct and immediate, and the symptoms that result from it are of a more spasmodic character. It is very easy to confound these two causes when they act simultaneously; but I have just shown that they can be considered separately. The following reason shows the importance of so doing:—

"Whenever we succeed in entirely putting a stop to the habits of abuse in children, we may make sure of obtaining their return to health, and that very quickly. This I have remarked in all the cases of children that have come under my care. I do not mean to infer that the disorder done to nutrition during the progess of development is easily repaired, but that the acute symptoms rapidly disappear, and that all the functions are quickly re-established. If the effects produced are active and serious they cease very rapidly, as soon as the cause is removed, and return to health becomes certain. Unfortunately matters do not follow so simple a course after puberty.

"What I have just said respecting children, applies equally to females; this is easily shown by examining the cases in which excision of the clitoris has been performed for the cure of nymphomania. The state of these unfortunates must have been deplorable indeed, to justify the resort to such means; yet they recovered very rapidly.

"Why in these two classes of cases, is the cure certain and the return to health rapid, as soon as the vice has been mastered? It is that the cause of the weakness immediately ceases to act on the economy. Why is it that so many men continue to waste away after they have entirely left off their habits of abuse? It is because diurnal pollutions have commenced, which are even more debilitating than the abuses which gave rise to them.

"Dr. Deslandes and many others have discovered that there is a great difference in the conditions of persons who have practiced masturbation for some time, and then renounced it; but they have not sought the explanation of this fact. It is, however, very important to know why some are cured rapidly and completely, while others continue to suffer and languish during the remainder of their lives. The symptoms experienced by the latter are those produced by diurnal pollutions. But if we inquire why some should be affected by diurnal pollutious while others are exempt, we discover that we have been comparing two very different classes of patients. The one class conquered their bad habits by the force of their will—the other class were compelled to renounce them by impotence. The former resisted their desires while they were yet active; they required much perseverance and moral energy in order to succeed. The latter only left off as they were less tempted—the progressive disease in their erections being due to the presence of undiscovered diurnal pollutions.

"Such patients deceive themselves as to the cause of their changing their habits, and are astonished at not finding any benefit arise from such change. Some of them even remark to their medical attendants that it is after they have left off their malpractices, that their health has become altered.

"All these circumstances, embarrassing at first sight, are easily explained on a little reflection. At first the genital organs are healthy; the constitution is uninjured; no seminal emissions occur except those that are induced voluntarily, and the activity of the digestive organs permits a rapid repair of the losses. But as soon as irritation is set up in the spermatic organs, a large quantity of semen is secreted, and escapes every day, and several times a day, without the patient's knowledge. The digestion is disordered; the erections and voluptuous sensations diminish, because the semen is less perfectly formed; the provocations are, therefore, weakened by degrees, and the patient renounces without difficulty habits which only inspire him with disgust. He wonders that his health still continues to grow worse, for he has not discovered that he passes daily, by often repeated evacuations, more semen than he formerly passed in a perceptible manner, and he does not take into account the difficulty felt by his economy of repairing these frequent discharges. We must not then confound those whose virility leaves them, with those whom the power of their will causes to recover, and we must not be surprised at seeing the alteration in the habits of each followed by very different consequences.

"In order to make the distinctive character of these two positions clear, I have laid stress on their most striking points. But there are numerous slight shades of distinction which I have not mentioned; for instance, in some cases the two classes of phenomenon occur successively in a very distinct manner, at very near periods. Many patients having corrected themselves once, find their health promptly re-established. But when, after recovering their strength, they have relapsed into their former habits, on renouncing them a second time, they obtain no benefit. These different results, under apparent similar circumstances, can only be explained by the occurrence of diurnal pollutions in consequence of the return to habits of abuse.

"Case 31 is a clear and perfect proof of the correctness of this explanation. The patient recovered twice after having twice conquered his passions; but the third time he only gave it up through disgust, and his health continued to deteriorate until cauterization arrested the diurnal pollutions from which he suffered.

"There are many circumstances which interfere with the good resolutions of those addicted to masturbation. After a few days of absolute continence, attained with much difficulty, they frequently suffer from nocturnal pollutions, the more frequent and the more abundant in proportion as the spermatic organs have been much irritated; the patients always feel more debilitated by these involuntary discharges than by those which they previously excited. Instead of combating these pollutions by suitable means, or after having employed one or two plans unsuccessfully, they think they will be able to diminish the evil by recurring to their former habits at distant intervals, and they thus relapse, increasing still more the irritation of the parts. Soon after diurnal pollutions commence and rapidly produce their effects; but as these are not discovered, the patients rejoice to find the nocturnal discharges gradually disappearing; but their health daily grows worse. This they cannot comprehend, and are frequently led to imagine that they have mistaken the cause of their disorder."

Many authors have noticed the indifference which persons, addicted to masturbation, show towards the opposite sex. This sentiment is, indeed, very common in those who have carried their abuses to a great extent; but I do not think it arises, as has been stated, from the long habit of solitary vice; at all events, I can assign a more direct cause for this indifference, viz.: the relative impotence of the patients—I say, relative impotence, because they possess sufficient power of erection to permit the practice of masturbation, but not enough to admit of sexual intercourse; and such patients seldom manifest any dislike to the opposite sex until they have experienced several disappointments, the remembrance of which constantly haunts them. Their views change immediately that the diurnal pollutions which kept up this impotence are arrested.

CHAPTER XIV.

EFFECTS OF TEMPERAMENT, IDIOSYNCRASY, &C.

The effects of abuses vary much in their characters and intensity, according to the individuals attacked. Some persons are uninjured by the most unbridled abuses, even when long continued, whilst others are very quickly disordered by slight abuse. In this respect I have witnessed very opposite cases with every variety of intermediate degree.

Temperament seems to have little influence in producing this inequality of resistance. Strength or feebleness of constitution is not of so much importance as might be supposed. The very unequal power of the genital organs affords the only satisfactory explanation. I shall refer to this point more fully when treating of venereal excesses.

Idiosyncrasy.—In the same individual, all the organs are not equally affected by abuse; this is shown by the frequent predominance of certain symptoms which give to the case a particular appearance, and are apt to lead to grave errors of diagnosis and treatment. I have related many cases in which this occurred. The presence of special symptoms, wherever a generally debilitating cause acts on the economy, arises from inequality of development or of activity, existing in certain organs. I shall at present only consider the direct and immediate action of abuses on the genital organs, so as to show the mode in which they produce nocturnal and diurnal pollutions.

CHAPTER XV

URETHRAL DISCHARGES.

Attacks of blennorrhagia (clap) are more frequent in persons addicted to masturbation than is generally supposed. Cases of this kind have frequently fallen under my notice. In the greater number of these patients the discharge was small in quantity, viscid and nearly transparent, or very slightly colored. It scarcely differed in appearance from the prostatic secretion; but in many patients the discharge was abundant, more or less colored, and attended with pain in the urethra, especially during the passage of urine. Several suffered from all the symptoms of a contagious blennorrhagia.

CASE XXXVII.

"In others the same symptoms recurred two or three times, and in one patient the discharges re-appeared as many as five times, always from the same cause. It is worthy of notice that there existed a kind of intermittance in the habits of the last mentioned patients; after having been moderate, or even quite continent for some time, they recommenced masturbation with fury, and the urethral discharges supervened on these relapses. Two of my other patients suffered from stricture of the urethra after one of these attacks of blennorrhagia, just as occurs after contagious blennorrhagia, and in one of these cases the stricture was very tight and very difficult of cure.

"I should remark that I am now speaking of patients who had never had sexual intercourse, and that I leave out of the question such as had suffered from cutaneous affections, in which the urethral mucous membrane might have participated. I must add that thirteen of such patients had not reached the age of puberty when the discharges occurred.

"These discharges, not having been excited by any virus, or by any constitutional disposition, must be referred to the effects of masturbation. Many of them having occurred before the age of puberty, it is evident that they could not consist of semen.

CASE XXXVIII.

PROSTATITIS.

"Several of my patients suffered from retention of urine, after the most frightful abuses; and it was necessary to relieve some of them with the catheter. In one patient an abscess formed in the prostate, and discharged through the pereneum.

CYSTITIS.

"I have related many cases of acute and chronic cystitis, of which masturbation was the sole cause.

EMISSIONS OF BLOOD.

"Some of my patients had carried their passions so far as to provoke emissions of pure blood, or of semen mixed with blood. Authors contain many such cases, which show that the pathological condition of the urethra has extended itself to the lining of the seminal vesicles. Other patients suffered from more or less severe attacks of hematuria; many experienced irritation of the bladder and kidneys, attended with an abundant secretion of bloody urine and constant desire to pass urine—sometimes even micturation was involuntary. Thus the inflammation or irritation caused by masturbation may, like that accompanying blennorrhagia, extend by degrees, until it reaches the kidney. It will be easily believed that the irritation does not extend in this direction only.

ORCHITIS.

"I have seen several cases in which the patients suffered from acute attacks of orchitis, after furious masturbation; and frequently such orchitis has required very active treatment for its relief. In one case the patient had not reached puberty when this occurred. In many such cases no doubt accessory circumstances existed, although the patients attributed the development of the orchitis only to masturbation. Others more slightly affected, experienced pain in the testicles and spermatic cords, accompanied with swelling of the epididymis. Others, again, suffered a painful sense of

.ension; they felt as if the testicles were held in a vice, or squeezed by a hand of iron. In many, the least contact of the parts with the clothes was insupportable, and the weight of the testicles caused very severe dragging pain. In all such cases the patients were obliged to wear suspensory bandages, and often to guard the testicles from friction with cotton, wool, or swans'-down.

"These symptoms which I have considered separately, generally occur together, and often form varying groups which present special appearances depending on the predominance of one of the symptoms. Sometimes the patients mention one circumstance only, because that one alone has attracted their attention; but when questioned, they recollect many others which appeared trifling by the side of the more serious one. It is also important to remark, that diurnal pollutions generally follow very soon after the appearance of these symptoms, and that the patients are a long time without discovering them, and sometimes only detect them when taught what to expect.

"The more we reflect on these morbid phenomena and the course of their appearance, the more striking is the resemblance between the effects of excessive masturbation and those of blennorrhagia. I admit that the symptoms do not always present the characters of well-marked inflammation, but they at least show those of active irritation of the parts. It is easy enough to give a clear explanation of what passes in all cases of this kind, with perhaps some slight shades of difference. The testicles secrete more semen, which is imperfectly formed; the seminal vesicles participating in the state of irritation of the neighboring organs, do not easily bear its presence—they contract more readily, as they are more easily affected by external impressions. Hence it becomes more and more difficult to avoid nocturnal pollutions. After a little time, diurnal pollutions occur and become more and more frequent and abundant; that is to say, there is a constant disposition in the seminal vesicles to contract spasmodically and expel their contents.

"On the other hand, the semen, ill secreted by the testicles, and remaining a shorter time in its reservoirs, becomes thinner and more watery; and by degrees, as it loses its physiological characters, it also loses its normal properties; it becomes, therefore, unfit to produce its effects on the seminal vesicles. The erections are consequently less energetic and less lasting, and after a time incomplete and fleeting, whilst, in the end, in severe cases, they disappear altogether. Hence the embarrassment and timidity of such patients in the company of females, and the fear they experience of finding themselves in a position to expose their impotence; and hence their indifference and even aversion for the sex, and the constantly increasing difficulty they experience in changing their habits. Such abuses, then, because their effects remain long after the habits have been altered, bring on symptoms, of which the cause is unsuspected. This is the reason why the health of some continues to deteriorate, whilst that of others, are re-established as soon as they have renounced their malpractices. This is why tonics, aphrodisiacs, cold bathing, and iced drinks, produce effects so different from those expected.

"There are undoubtedly cases, in which the spermatic organs are weakened and relaxed. I shall relate several instances of this in a future chapter; but we shall then see that such a state arises from primary relaxation of habit and rather from want of use of the organs, than from their abuse.

"In concluding my remarks on the subject of masturbation, I may observe, that it is the most dangerous of all vices of this nature, because it is the most difficult to discover and to prevent, and because it does not require any assistance for its consummation. From the cases I have seen, I conclude that the irritation excited by such manœuvres very easily induces involuntary discharges; that the appearance of nocturnal pollutions in those who attempt to abandon the vice, often causes them to return to their former habits, and that the diminution of virility which follows, far from favoring the patient's amendment, frequently hinders it by proving an obstacle to their having sexual intercourse, while it does not prevent them from continuing their bad practices. This circumstance is a powerful cause of the disorders which attend such as are reduced to vicious habits."

CHAPTER XVI.

EROTOMANIA AND SATYRIASIS.

These affections are generally confounded together, yet the difference is very considerable, though their manifestations are similar. There is an unnatural sexual desire in both of them, which is often so unconsolable, that gratification is attempted at any sacrifice. They are produced by disease, as has been shown in the preceding Chapters of this work.

Satyriasis is caused by a disease of the sexual organs, or the adjacent parts, which keep them in a constant irritation, often so great as to prevent proper sleep and causing the greatest uneasiness at all times, producing a most furious excitement; those which are most likely to produce satyriasis are the piles, ascarides, bladder, and the urethra or prostate gland. Dr. Curling remarks, that

"The irritation attending the morbid condition of the mucous membrane of the prostatic portions of the urethra, tends in a very material degree to excite both the excessive seminal discharge and the secretions of the prostate, and to produce that morbid craving for indulgence and abuse which persons who have brought themselves to this state find so difficult to repress and resist.

"It is well known that any irritation at the orifice of an excretory duct usually acts as a stimulus to the secretion of the gland. Thus, hurtful matter in the duodenum produces a flow of bile; and a foreign body in the conjunctiva, as an inverted eye-lash, a discharge of tears; so it is with the testes when irritation exists at the orifices of their excretory ducts. The disorder at this part, moreover, appears to react on the brain, and to become in part the cause of the patient's mind being constantly occupied with subjects of sexual excitement, and of his indifference and apathy to other matters; so that the local disease induced by abuse powerfully aids in perpetuating the mischief; and, judging from the experience which I have had in these cases, is the object to which our treatment should be first directed.

"In all of these cases reasoning is of no use whatever; the cause must be removed or the disease will continue for the patient can no more control himself than he could if he was laboring under a diarrhœa or fever.

"This is a truth, I fear, not sufficiently impressed on the minds of medical men. One would be loath to offer any apology for the vicious habits and indulgences to which, it is well known, old men are occasionally addicted—a melancholy example of the kind, in the higher ranks of life, having lately been brought under public notice. I cannot but think, however, that, in many instances, these cases are not undeserving of professional sympathy, and that the erotic longings—which sometimes continue to distress the aged long after the period at which, in the course of nature, they should have ceased—depend as much on physical infirmity as mental depravity, the former inciting and producing the morbid desires. If these propensities were regarded and treated as symptoms of disease, (and that they frequently occur in connection with affection of the urinary passage, is well known to practical surgeons), I believe they would often subside, and the distressing results to which they lead would be altogether avoided."

Erotomania differs from satyriasis as respects the seat of the disease, as the former is in the brain and not in the genitals directly, for they are only affected in a secondary manner.

Dr Copeland is very clear in his distinction of the two diseases; but I will here say, nymphomania, to which he refers, is the same disease in the female as satyriasis in the male. He says,

"*Erotomania—Monomanie erotique of Esquirol*, is characterized by an excessive love of some object, real or imaginary. It is a mental affection in which amorous ideas are as fixed and dominant as religious ideas are in religious momomania or melancholia. Erotomania is very different from satyriasis and nymphomania. In the latter the mischief is in the reproductive organs; in the former it is in the mind. The one is a physical, the other a moral disorder.

"Erotomania is the result of an excited imagination, unrestrained by the powers of the understanding; satyriasis and nymphomania proceed from the local irritation of the sexual organs, reacting upon the brain, and exciting the passions beyond the restraints of rea-

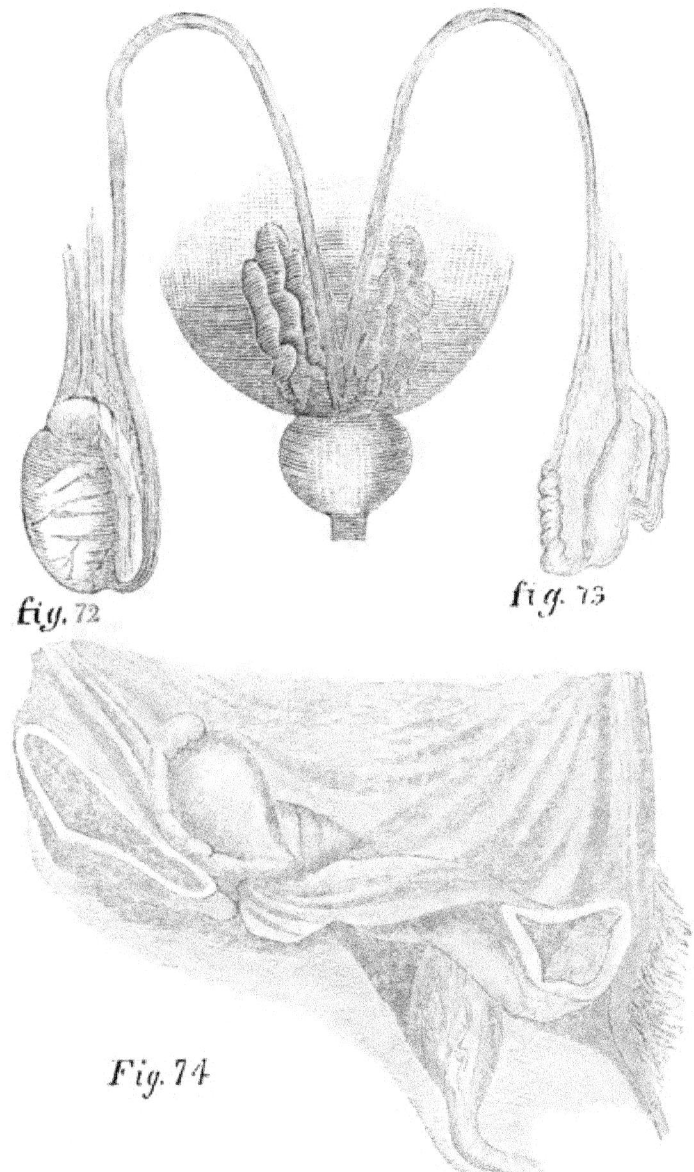
fig. 72
fig. 73
Fig. 74

son. In the former, there is neither indecency nor the want of chastity; in the latter, there is unrestrained expressions of sexual desire and excitement. The one is commonly caused by ungratified or disappointed affection excited in a virtuous mind; the other, by inordinate irritation or indulgence of the sexual passion.

"In erotomania, the eyes are bright, the manner and expressions tender and passionate, and the actions free, without passing the limits of decency. Self and selfish interests are all forgotten in the devotion paid, often in secret, to the object of the mind's adoration. A state of ecstasy often occurs in the contemplation of the perfections which the imagination attaches to the subject of its admiration. The bodily functions languish during this state of moral disorder; the countenance becomes pale and depressed; the features shrunk; the body emaciated; the temper unquiet and irritable; and the mind agitated and despairing. The ideas continually revert to the loved and desired object; and opposition or endeavors to turn them in a different direction only render them more concentrated and determined in their devotion. At last, parents and fortune are abandoned, social ties broken asunder, and the most painful difficulties are encountered, in order to obtain the object of admiration.

"In some cases the attempts made by the patient to conceal and overcome this affection, occasion a state of irritative fever, with sadness, depression, loss of appetite, emaciation, &c., which has not inappropriately been termed by Lorrey, Erotic Fever, and which, after continuing an intermediate period, may even terminate fatally.

"When a young person becomes sad, absent in mind, pale and emaciated, sighs frequently, sheds tears without any obvious reason, is incapable of any mental or bodily exertion, scarcely speaks to any one, loses appetite, &c., it is sufficiently evident that the mind is inordinately possessed by some desired object. If a strong effort be not made to dispossess it of the predominant sentiments, or if the object of desire be not obtained, the symptoms become still more distressing. The corporeal functions languish, the eyes sink, the pulse becomes weak and irregular, and the nights disturbed and sleepless. At last a form of slow hectic is produced, and the weaker organs, especially the lungs and heart, are the seat of slowly-produced disease; the whole frame is blighted, and the patient sinks from the injurious influence of the mental affection on the vital organs

"This form of moral disorder may increase, and affect the intellect in a much more serious manner, until general insanity or mania is developed; and, with the progress of time, it may at last terminate in dementia or incoherent insanity. In each of these, the primary character of the disorder, or the original moral affection, will still continue to be manifested by the frequent suggestion of the same train of ideas, or recurrence to the object of devotion."

I need not impress the importance of the selection of a proper physician in all of the cases, which I have so fully treated on, in the foregoing pages, for I think the facts therein presented, and the immense deal of suffering and fatality, which has and still is caused by the physicians not understanding the proper treatment—from the lack of the necessary practice—will put all who are unfortunately afflicted, or who have any friends or relatives in such a position, on their proper guard, and where the most worthy among so great a number—saying nothing of their fitness to be employed—can be found.

CHAPTER XVII.

THE INFLUENCE WHICH DIFFERENT ARTICLES HAVE—MEDICINES AND OTHERS—IN PRODUCING AND CURING SPERMATORRHŒA AND IMPOTENCE.

There are not many medicinal substances that act directly on the genital organs; some of those that do so, are beneficial, and some of them highly injurious. From the immense deal of imposition practiced both by the quacks and regular practitioners, I have concluded to present a few remarks and extracts, that others may escape their base productions, and thus save the consequent injury that their use would be sure to cause.

There has always been an ignorant notion existing among the people generally, that physicians were acquainted with the secret specification of certain medicines on the generative functions, which would arouse a sexual ardor under almost any circumstances, and in order to play upon the credulity of the public, and at the same time better their pecuniary circumstances, have originated the various cordials, elixirs, stimulants, &c., which are so much advertised, in order to attract the attention of those who are troubled with emissions, seminal weakness, impotency and sterility. None of them are, in the slightest degree, capable of performing what is promised of them.

It is only very recent that a discovery has been made of certain drugs which will produce the most astonishing beneficial action, in those cases suffering from loss of power, yet each individual case requires its being administered with a skillful hand, after a full statement of the case has been presented, and it is to be hoped the knowledge of the ingredients will be kept from those charlatans who only know how to prescribe for all alike, notwithstanding the differences presented in almost every case.

Purgatives are thought by almost everybody to act only upon the bowels, but instead of that, as I have already shown, irritation from spasmodic contractions of the rectum extends to the seminal vesicles, and produce just as great and distressing diurnal pollutions as those which arise from mechanical compression of the same organs; so that of ascarides, diarrhœa, &c., produce emissions, they as a matter of course will be caused by drastic purgatives, because they act chiefly by irritating the large intestine, and which bring on in the rectum and neighboring parts a more permanent state of irritation, and consequently excite an unnatural flow of urine by irritating the bladder. Now, the spermatic ducts and prostate gland are close to the rectum, and are just as much affected as the bladder—which is one reason why patients are injured instead of being benefitted by using such a mode of treatment—because they are costive, which they generally are. I have had a great many patients from physicians who so foolishly adopted such treatment, in a worse state than they were before they visited them.

NARCOTICS.

I have often been obliged to prohibit patients the use of tobacco, liquor, coffee and tea, and as evidence that other surgeons also have found some or all of them injurious I will here insert some of their remarks.

Coffee—M. Lallemand says, affects the cerebro-spinal system, and that attention has not been paid to its action on other organs. Taken in moderate quantities, it excites the bladder and kidneys, increases the secretion of urine, and renders its discharge more frequent. It acts in the same manner on the spermatic organs, augments the venereal desires, favors erections, and accelerates ejaculation. Taken in excess, however, it produces injurious effects.

Tea.—M. Lallemand is supported in his views of the action of tea by Dr. McDougal, and is satisfied its effects are very similar to coffee. He further says, there are many other agents besides those he has mentioned, which excite or increase involuntary seminal discharges.

Tobacco.—This article produces similar effects on a great many persons to opium and other narcotics. At first they often stimulate, but afterwards they weaken the sexual organs, so much so as to cause complete impotency. This no doubt is caused by the relaxation of the ejaculatory ducts.

In the thirteenth Annual Report of the Massachusetts State Lunatic Asylum, there are excellent remarks on tobacco, which I cannot refrain from inserting. After the remarks on the injurious effects of alcohol it goes on to say,—

"Alcohol is not the only narcotic which thus affects the brain and nervous system. Opium produces delirium tremens, and probably insanity. Tobacco is a powerful narcotic agent, and its use is very deleterious to the nervous system, producing tremors, vertigo, faintness, palpitation of the heart, and other serious diseases. That tobacco certainly produces insanity, I am not able positively to observe; but that it produces a predisposition to it, I am fully confident. Its influence on the brain and nervous system generally, is hardly less obvious than that of alcohol, and, if used excessively, is equally injurious. The young are particularly susceptible to the influence of these narcotics. If a young man becomes intemperate before he is twenty years of age, he rarely lives to thirty. If a young man uses tobacco while the system is greatly susceptible to its influence, he will not be likely to escape injurious effects that will be developed sooner or later, and both diminish the enjoyments of life and shorten its period.

"The very general use of tobacco among young men at the present day is alarming, and shows the ignorance and devotion of the devotees of this dangerous practice to one of the most virulent poisons of the vegetable world. The testimony of medical men, of the most respectable character, could be quoted to any extent to sustain these views of the deleterious influence of this dangerous narcotic.

"Dr. Rush says of tobacco, 'It impairs the appetite, produces dyspepsia, tremors, vertigo, headache and epilepsy. It injures the voice, destroys the teeth, and imparts to the complexion a disagreeable, dusky brown.'

"Dr. Boerhave says, that, 'Since the use of tobacco has been so general in Europe, the number of hypochondrical and consumptive complaints have increased by its use.'

"Dr. Cullen says, 'I have known a small quantity snuffed up the nose to produce giddiness, stupor and vomiting. There are many instances of its more violent effects, even of its proving a mortal poison.'

"Dr. Darwin says, 'It produces disease of the salivary glands and the pancreas, and injures the power of digestion by occasioning the person to spit off the saliva which he ought to swallow.'

"Dr. Tissott once saw the smoking of it prove fatal.

CASE XXXIX.

"Dr. Pilcher details the particulars of a case of a medical student whom he had been requested to see 'This gentleman suffered under all the symptoms of phthisis. There was muco-purulent expectoration night sweats, &c. The mucous membrane of the throat, epiglottis, and the neighboring parts, were coated with a brown fur. The patient had been an immoderate snuff-taker; he was told to discontinue the snuff—he did so, and recovered.'

CASE XL.

"Dr. Chapman says, 'By a member of Congress from he west, in the meridian of life, and of a very stout frame, I was some time since consulted. He told me, that from having been one of the most healthy and fearless of men he had become sick all over and timid as a girl. He could not even present a petition to Congress, much less say a word concerning it, though he had long been a practicing lawyer, and served much in legislative bodies. By any ordinary noise he was startled or thrown into tremulousness, and afraid to be alone at night. His appetite and digestion were gone; he had painful sensations at the pit of the stomach, and unrelenting constipated bowels. During the narrative of his suffering, his aspect approached the haggard wildness of mental distemperature. On inquiry, I found that his consumption of tobacco was almost incredible, by chewing, snuffing and smoking. Being satisfied that all his misery arose from this poisonous weed, its use was discontinued, and in a few weeks he entirely recovered.

CASES XLI AND XLII.

"Distressing as was this case, I have seen others, from the same cause even more deplorable. Two young men were in succession brought to me for advice, whom I found in a state of insanity, very much resembling delirium tremens. Each had chewed and smoked tobacco to excess, though perfectly temperate as regarded drink. The further account given me was, 'that in early life adopting this bad practice, it grew with their growth. Dyspepsia soon occurred, attended by great derangement of the nervous system, and ultimately the mania, I have mentioned; but I have also seen the same condition very speedily induced.'

"Dr. Franklin says, 'He never used it, and never met with a man who did use it that advised him to follow his example.'

"The venerable John Quincy Adams, in a recent letter on the subject, says, that in early life he used tobacco, but for more than thirty years he had discontinued the practice. 'I have often wished,' says he, 'that every individual of the human race, affected with this artificial passion, would prevail upon himself to try but for three months, the experiment which I have made, and I am sure it would turn every acre of tobacco-land into a wheat-field, and add five years to the average of human life.'

"Some cases have come under my observation which show the injurious effects of tobacco where no evil was suspected.

CASE XLIII.

"A respectable merchant, who abstained wholly from ardent spirits, applied to me for advice. He complained of great weakness, tremor of the limbs and joints, with lassitude, general prostration of health, and depression of spirits. Knowing that he used tobacco freely, I advised him to discontinue it entirely; he soon became better, and after a time was wholly relieved from these disagreeable symptoms.

CASE XLIV.

"A distinguished clergyman informed me that he had been an extravagant snuff-taker; that for years he had a disagreeable affection of the head, and his health was not good. He did not attribute either to the use of snuff, but thinking it a filthy habit and a growing evil, he resolved to leave it off. He was surprised to find that the difficulty in his head almost immediately left him, and his general health became quite good.

CASE XLV.

"A gentleman of athletic frame, and about twenty-four years of age, applied to me for advice. He complained of insufferable faintness and distress of stomach morning sickness, vomiting, trembling, and prostration of strength. He diminished his tobacco considerably, and was immediately better, but had not resolution to abandon the pernicious practice.

"In our experience in the Hospital, tobacco in all its forms is injurious to the insane. It increases excitement of the nervous system in many cases, deranges the stomach, and produces vertigo, tremors, and stupor in others. It is difficult to control its use with the insane, and although considerable suffering comes from its entire abandonment, it cannot be generally allowed with safety

CASE XLVI.

"One patient, while at labor, found a quantity of tobacco, and hid it in his bed. He used it freely, became sick, lost his appetite, and confined himself to his bed, completely intoxicated. After some days, diligent search was made, and a store of tobacco was found in his straw bed; when this was removed he almost immediately recovered, and in a few days he was as well as before.

CASE XLVII.

"A person who came into the Hospital a furious maniac, soon became calm, and improved favorably. He labored in the field with propriety, and exhibited every indication of a favorable convalescence. Suddenly without any apparent cause, he again became very violent and insane. It was soon discovered that he had in some way obtained tobacco. After he ceased to use it, he again became calm and convalescent.

CASE XLVIII.

"An aged lady was brought to us very insane. The practice of her friends for some time, had been to give her ardent spirits to intoxicate her at night, and tobacco and snuff, in unlimited quantity, for the day. All these were withdrawn at once; her sufferings for some days were great, but after a time, she became calm, and got better as soon as the influence of this excitement was over.

"It is very natural to suppose that an article possessing the active properties of this fascinating narcotic, should produce most deleterious effects upon health—particularly upon the brain and nervous system.

"The uninitiated cannot smoke a cigar or use tobacco in any form, without unpleasant effects—how then can it be possible that a poison so active can be used with impunity? The stomach and brain, subjected to such influences, will become diseased, and show their effects as certainly as if alcohol were used.

"If asked my medical opinion, which was safest, four glasses of wine or four quids of tobacco, daily? I should say, unhesitatingly, the wine. Of the two evils, this would, in my opinion, be the least. Tobacco is the strongest, most dangerous narcotic—the habit of its use is the strongest and most difficult to overcome, and the influence felt from it most baneful and destructive to health."

Opium, of course, is much worse than tobacco, as the exhaustion it produces is not so readily recovered from, for if its use once cause impotency, it is quite doubtful if he can be restored to his former powers.

I will close this article by quoting what M. Lallemand says, in a number of his cases, on its effects in decreasing and otherwise affecting the sexual desires. I very often have patients affected in the same way.

CASE XLIX.

"I have a young man of very nervous temperament at present under my care, in whom nocturnal and diurnal pollutions have brought on pain in the loins, palpitation, difficulty of breathing, &c., symptoms which were supposed to arise from disease of the spinal cord, cardiac affection and commencing phthisis. Among the exciting causes of these involuntary discharges, the effects of smoking occupy the chief place. The following is the patient's statement:—

"'At twenty years of age I wished to accustom myself to smoking; but a day never passed without my experiencing complete intoxication, attended with vomiting, vertigo, and trembling of the limbs. I continued the habit, however, and I soon began to perceive that my sight became weak, and that I lost my memory; my hands shook, and my digestion became much disordered. I noticed also great debility of the genital organs; my erections ceased; and at the age of twenty-two I found myself completely impotent.'

"This patient had rarely practiced masturbation, and had never committed any excess when he first began to smoke; his health had previously been excellent. It is, therefore, evident, that the impotence as well as the other symptoms, arose from tobacco. Impotence at the age of twenty-two can only be produced by involuntary seminal discharges, provided there be no physical disability. In the present case, there was no doubt on the point, the patient himself having discovered diurnal and nocturnal pollutions.

"The action of tobacco on those who smoke for the first time, is too well known to require description; more or less disorder of all the functions, varying according to the constitution of the individual, invariably arises from it; and this disorder always presents more or less of the characteristics of poisoning by narcotics. These effects go off by degrees, as the patient becomes habituated to the use of tobacco, and generally after a time cease to be manifested at all. Some nervous and excitable individuals are unable to accustom themselves to the habit, as in the case just mentioned; in others, again, smoking becomes an artificial habit, which in many cases is almost a necessity.

"But this empire of custom has its limits, beyond which the narcotic influence re-appears. In such as are not easily affected, this acquired habit is generally supported with impunity; but even then, if it be indulged in to excess, it must after a time be injurious. Thus it is that the most accomplished smokers often experience vertigo, cephalagia, anorexia, &c., when they have remained long in an atmosphere densely filled with smoke, which is then drawn into the lungs, and probably produces worse effects than when merely drawn into the mouth, or swallowed, as in smoking.

"In a word then, if the power of habit can prevent the momentary effects of smoking from showing themselves, the frequent repetition of the use of tobacco produces more lasting effects on different organs. Disorder of the digestive organs is well-known as occurring in inveterate smokers—that of the genital organs has not hitherto been noticed."

Dr. McDonald says, that "Many inveterate smokers among his professional friends have mentioned to him the diminution of their veneral desires, as one of the effects of tobacco."

CANTHARIDES.

The application of a blister on a particular part of the body will not only increase involuntary seminal discharges, but will cause them. Cantharides will do this, whether used as a blister or administered as an internal remedy, because it will cause an inflammation of the bladder and the contiguous parts of the body to such an extent as to prevent the passage of the urine. This is the reason that physicians have administered this drug in cases of seminal discharges and impotency, supposing them to be caused from atony or a relaxation of the genital organs. But of the hundreds that have come to me, after its use, not one had received the slightest benefit; but, on the other hand, had been injured to such an extent as to be troubled with a constant bearing down and other pains in the region of the bladder, obliging them to pass urine (in one bad case,) as often as ninety times within the twenty-four hours. Many patients received an increase of all their symptoms, and impotence taking the place of the weakness of the genitals. Injecting the urethra produces the same result.

If a physician does not understand the nature of a disease, I am surprised that he would have even the remotest idea of undertaking to cure it, much more doing so, for if they did understand it, they would not use cantharides, for instead, as they suppose, no doubt, of the disease—Emissions—being caused by a weakness of the organs, they are produced by irritation, and will increase instead of being removed, unless that irritation and cause is removed, which can never be done by this nor any other medicines; yet, see patients every day that have been under the care of physicians, or used all of the remedies which are so much lauded in the papers as being the only cure.

The public have long labored under the great error of supposing cantharides would create or restore power to the genital organs, and a desire for sexual gratification, to an extent that would satisfy the purposes not only of those who really needed something of the kind ; but it has been given for base purposes in many a case, I have no doubt. Many a female case has been presented to me of great sufferings, caused by its use.

There has been very recent discoveries of harmless medicines, which are surprisingly effective in cases of impotency, sterility, and weakness of the generative system ; but I administer it, of course, in accordance to the peculiarities of each case.

CAMPHOR.

This is a good antidote for the injurious consequences of the use of cantharides ; but its use for the cure of spermatorrhœa often causes injury.

CASE L.

As an instance of its injurious effects, M. Lallemand says, " One of his patients who put camphor between the prepuce and glans penis, suffered so much from diurnal pollution as to endanger his life."

It is a drug which quacks use, and I have found in a great many patients, after having used their remedies, an augmentation instead of diminution of their symptoms

NITRATE OF POTASS.

I consider this almost as injurious and dangerous as cantharides ; but notwithstanding all its injurious effects, almost all doctors, and nearly every quack remedy, especially for the cure of gonorrhœa, gleet, &c., incorporate it with their other drugs. In the inflammatory stage of these diseases it is supposed to have a sedative effect. It is considered a diuretic, because it increases the flow of urine, thereby stimulating the kidneys. But if it would stop there it would not cause so much harm, for it produces hematuria, pain, and inflammation of the bladder, consequently swelled testicles, and increase of, or will originate an inflammation of the urethra and prostate gland. As further proof of its injurious effects I will insert a case mentioned by M. Lallemand :—

CASE LI.

"A merchant of Gênes, wishing to take a purgative, sent to a druggist for an ounce of sulphate of magnesia. By mistake an ounce of nitrate of potass was returned by the messenger and taken. Violent inflammation of the urinary passages, accompanied with a discharge resembling blennorrhagia (clap) resulted, swelling took place in about the centre of the urethra, and when the acute stage of inflammation had passed off, a circumscribed induration, which obstructed the discharge of urine, remained. Twenty years afterwards the patient still suffered from this obstruction, for the formation of which there had been no other cause than the inflammation produced by the nitrate of potass. The patient had never had blennorrhagia, either before

"It appears then the nitrate of potass acts as a stimulant of the whole urinary apparatus; and it is at least probable that it produces the same effect on the spermatic organs. I am led to this opinion partly by analogy, but chiefly because more than forty of the patients whom I have treated for involuntary seminal discharges, had taken nitrate of potass in some form or other; and all, without exception, found themselves worse afterwards. Many of them also observed the same effects from preparation of squill, and, in fact, or after, and had never suffered any injury of the part. all other diuretics."

ERGOT OF RYE.

This article is generally used by physicians to expedite delivery in females, by increasing the action of the womb. It also acts as a stimulant to the male genitals, by creating an inflammation of them, which of course produces seminal emissions. It is a dangerous and highly injurious drug.

PHOSPHORUS

This is the most dangerous of all the articles I have yet mentioned, and therefore should never be thought of for a moment, especially by those who are not perfectly well acquainted with its action. Its effects are similar to cantharides, only a great deal more energetic, which renders it so much more dangerous, for its injuries are not easily recovered from

IRON.

There are various preparations of this mineral, which have been and still are used by a great many physicians in the treatment of spermatorrhœa. In what respect they consider it beneficial, I do not know, unless as a tonic; that is what it is given for, no doubt. As a tonic for the system, it is useful in the proper cases, but as to any special benefits, or any particular influence directly on the genital organs it possesses, I think is very indefinite and unsatisfactory indeed. Almost every patient who comes to consult me, after having been under previous treatment, has used one or the other of its different preparations.

CHAPTER XVIII.

THE MECHANICAL MEANS ADOPTED FOR THE PREVENTION OF SEMINAL EMISSIONS.

The reader will have noticed what some of my patients experienced, by compressing the urethra at different places, from in front of the testicles, back as far as the anus, for the purpose of preventing an erection of the penis and the semen from being ejected. I have also copied cases from M. Lallemand, where the same means had been used by his patients. Fournier and Begin both had patients who had adopted the same means to prevent the semen from escaping, thereby, as they thought, effecting a cure. The different means adopted to attain this end, I will now mention.

CASE LII.

A few years ago I had a patient, aged about thirty, who had, among the rest of his numerous successful remedies, adopted this one, and for a long time he supposed he had rid himself entirely of the harassing and dreaded nocturnal pollutions. At night, on going to bed, he would wind a piece of tape losely around the penis while it was in its natural state. This would cause pain, when the penis became erect during the night—which it did at that time when an emission took place—and wake him from his sleep. For a long time he did not see any semen escape from the urethra, and he flattered himself—though he had grown weaker and thinner—he had hit upon a very simple but sure cure. One morning, however, he discovered in the bottom of the vessel a thick mattery substance, which led him to examine the urethra, when he discovered what he was satisfied was semen, gradually oozing out, after he had passed urine. By examining the urethra after stool, he discovered the same thing, and, as his erections had ceased for some time, as well as all feelings of venereal excitement, he knew the semen was gradually escaping almost constantly. The physicians he had before consulted, not benefiting him in the least up to the time he applied his own remedy, he very wisely came to the conclusion not again to put himself under their charge. This was his situation when he applied to me. In a few weeks I cured him of his pollutions, and in a year after, he was a married man, in all the enjoyment of good health.

A great many of my patients have adopted different means for the attainment of the same object--that is, the escape of semen. Some have applied pressure on the pereneum by means of a pad; some have placed grooved pieces of wood on each side of the penis, with tape wound around it; others again have procured rings to fit the penis; but the result in every case has been to cause the semen to flow into the bladder instead of out, through the urethra, which would not have caused half as much injury as resulted from the former.

In Case 37 of M. Lallemand's work, the patient ruptured the urethra by compressing it to prevent the discharge of semen, and which was followed the following day by a discharge resembling a gonorrhœa, and which he only succeeded in curing by cauterizing the canal. In this case this one compression created so much irritation as to immediately cause involuntary emissions. The strangulation or compression of the urethra causes so much irritation at the orifices of the seminal ducts that it produces an inflammation in that part of the canal which extends itself either way, and, finally, if the practice is continued for only a short time even, will cause a stricture, (and in some cases I have found as many as three in different parts of the canal which were irritable and otherwise very bad), in addition to the diurnal pollutions, an inflammation of the prostate gland.

CASE LIII.

In one of my cases, a man of about forty years of age, was so severe that it ulcerated and endangered his life. The bladder is sure to become involved also, as will be seen, when you consider the amount of the foreign substance that is being constantly thrown into it The inflammation of the bladder, in these cases, is often of the severest nature, and sometimes requires a long time, with the most energetic treatment, to restore it to its natural state.

Every human being, with an intellect sufficient to enable them to distinguish between right and wrong, will at once brand those who continue to use such mechanical treatment, in defiance, and regardless of such undeniable proofs, of the certain injury to those persons who may be so unfortunate as to accidentally adopt their mode of treatment, as arrant knaves and impostors, or ignoramuses, that would do mankind a better turn by street-sweeping than the practice of an honorable profession. Such, however, I am sorry to say, are base enough to advertise appliances of the above description, as great inventions, and the only thing that is capable of effecting a cure. One impostor to more easily mislead the ignorant, says, "the medical faculties in Europe use them." The assertion is a base falsehood; the public will rightly judge such a character.

When any person is so unscrupulous as to say, that there is no one else but himself capable of effecting a cure, of whatever complaint he may mention, it seems to me the public would be unanimous in condemning him as an impostor, for no one but a knave would have the effrontery to arrogate to himself such superiority over all others. There is a falsehood on the very face of all such declarations, which humanity at this enlightened day should be able at once to discover. The treatment, remedies, of all such persons, cause a thousand times more injury than the disease itself, as I think these pages sufficiently prove.

M. Lallemand says, "Anti-venereal treatment is frequently also employed for patients who have suffered merely from blennorrhagia, and in a very numerous class of cases it produces a serious increase of the irritation in the genital organs, and causes the appearance or exasperates the effects of involuntary spermatic discharges." Cases of this nature often present considerable difficulties of diagnosis; and the solution of these obscurities is always of much importance in determining the treatment to be followed.

Anti-venereals are not the only therapeutic agents which produce such unfortunate effects; those which a blind routine of practice employs in cases of blennorrhagia have not been less injurious. Among these it is especially necessary for me to mention astringent injections, copaiba, cubebs, tonics, and bitters, employed too soon, or in extreme doses. All these means act more or less by exciting the genito-urinary organs; it is, therefore, easy to understand that their untimely or immoderate use must favor an extension of the inflammation from the urethra to the mucous membranes, which are continuous with it.

"Lastly, spermatorrhœa is often made worse by the very means employed for its removal, and among these may be ranked cold baths, ice, tonics, bitters, sulphur-baths, &c.

"The further we advance the more plainly we shall

see how necessary it is for the different forms of spermatorrhœa to be described as simple affections—how necessary it is to regard them in all their aspects, and to take account of all the circumstances which assist in producing them. In practice we find it indispensable to weigh well all the points connected with a case of spermatorrhœa, before deciding on any diagnosis, prognosis, or especially on our treatment."

CHAPTER XIX.

NYMPHOMANIA PRODUCTIVE OF MONOMANIA AND CRIME—MASTURBATION PRACTICED BY FEMALES—REMARKS SHOWING ITS DREADFUL EFFECTS ON THE CONSTITUTION.

CASE LIV.

The following extraordinary case of Nymphomania was related to me by one of the most prominent physicians of this city:—

Miss ——, of a wealthy and highly respectable family, residing in one of the thousand-palaces of —— avenue, attracted my attention while engaged in my professional attentions to a member of the family, by marked respect at first; soon, however, her manner changed to sociable familiarity. Yet I only regarded this as a clever display of a brilliant mind, highly educated. Subsequently, the pathetic execution of several love songs, sung to me while, apparently by accident, alone with her, led me to observe her conduct narrowly, which the lingering illness of my patient gave me ample opportunity to do. By closely scrutinizing her conduct, I discovered every possible means resorted to, to secure my company and attention. Thus, from day to day I observed the development of criminal passions, in this beautiful creature. The slightest accident soon directed my attention to the real cause, not until, however, after frequent and extravagant demonstrations of a criminal passion.

heated by an ardent temperament, perpetrated the warmest and most eloquent declaration of love.

The affair now assumed an alarming aspect. I being her senior and a married man, I reasoned, remonstrated, threatened to disclose the whole conduct to her parents, which only fanned the passions to a blaze.

The accidental discovery of the cause occurred in this way:—In going to my patient's sick-room I passed, one warm evening, the bath-room, the door of which stood ajar; a glance revealed the secret of this strange conduct. I saw this poor creature sitting in a bath, which her mother told me that she too frequently used as a cold sitting-bath, asking my opinion of the propriety.

I observed, after the bath, her conduct was invariably modest. The cause of this strange conduct then was inflammation and excessive irritation of the genital organs.

After a consultation with the parents, I explained to her, her situation, which I was always compelled to do, immediately after the cold hip-bath, to prevent the immodest conduct of the absolutely insane young lady.

This then, sir, is the case for which I wrote you the note asking your plan of treatment, which, I am happy to acknowledge, speedily restored my unfortunate patient to perfect health, and who is so grateful to you, as I deemed it my duty to tell her confidentially that you was the savior of her virtue, of her health, and of her happiness.

She begs you will accept the accompanying $100 note, with the wish that you will give her the opportunity to make her personal acknowledgments.

Yours, truly, M. D.,
Bleecker st., N. Y

New York, June 11, '52.

Dr. Larmont:

Dear Sir,—In answer to your request of the publication in your popular work, of the interesting case of nymphomania, for which I consulted you, I am happy to inform you that the young lady feels so greatly indebted, that for the sake of humanity and in consideration of your invaluable services, she cheerfully consents to your request.

It is not necessary, of course, to ask to conceal everything that would lead to suspicion who our patient is, as the delicacy of a charming, talented and admired young lady, would suffer the deepest mortification, although her conduct was the direct consequence of physical disease. Accept my kindest regards,

Yours, truly, M. D.,
Bleecker st., N. Y.

This case is one only of a number which has passed under my own knowledge. The complaint produced what we might call a species of monomania. Whether the ardent temperament! an over-healthy or very vigorous constitution! ascarides or worms in the rectum! or masturbation, is the cause, the result is the same. Such—yes, many cases like this, do exist, and parents and guardians should be able to obviate a criminal result, by arranging a proper marriage; or, if produced by any foreign excitant, as masturbation, or the troublesome and tormenting little worms, the experienced physician in the cure of these complaints, should be consulted, without a day's delay. By adopting these timely remedies, they may not only save their child or relative from an ignominious and horrible end, but themselves from everlasting disgrace.

Onanism particularly exists among females; and that too, to a very alarming extent. It is scarcely possible for an unprofessional person to conceive the long train of ills it produces. It is a knowledge of this fact, as I have stated in the preface to this work, that induced me to present to the world truths which will serve as a saving monitor to the unsuspecting of both sexes.

We address these few pages to mothers and guardians, with the two-fold view of furnishing them with the means of preserving the morals of their daughters, and of sparing them the pain and sorrow of seeing them wither and perish at an age when they ought to be the ornament of their domestic circle, and enjoy health and happiness. This life-destroying habit, will enable parents and guardians of youth, to recognize by numberless symptoms, the gradual disease which is hurrying them to the tomb; and I must reiterate a previous assertion, that I cannot but hope this work will be the means of saving many from a lingering suicide. Let every mother, father, and guardian, therefore, read this work, and they can learn from it, whether their children or charges are safe from this vice, or have strayed from the paths of health and chastity.

Have you ever seen a lovely flower, when the least breath would scatter its leaves to the winds of heaven, and yet retain its original loveliness? Such a flower is the innocent child which is on a brink so perilous, that, unless warned at once, is sure to perish and leave sorrowing relatives to mourn its loss.

Health and beauty are not the only blessings to be preserved! All moral feeling, all proper sentiments, the happiest gift of intellect, and every hope of happiness will surely be destroyed.

These individuals who yield themselves up to the enjoyment of solitary pleasure, to secret pernicious habits, soon exhibit more or less the symptoms of tabes dorsalis —a species of consumption. At first they are not troubled with fever; and, although they may still preserve their appetite, their bodies waste away; walking particularly wearies them and produces profuse perspiration, headache, ringing in the ears, derangement of the nerves and brain, and terminating with stupidity. The stomach becomes deranged, the patient is pale, dull and indolent; their eyes are hollowed, their bodies thin, their legs can no longer sustain them; they are totally unhinged, incapable of any exertion, and in many cases attacked with palsy. The weakness and constitutional injury thus produced are the reasons why such patients bear less easily those diseases which all are subject to. The chest particularly becomes affected, indigestion comes on, the most robust girls are soon rendered weak; and sometimes a slow fever, a rapid consumption, or apoplexy, soon terminates the scene.

Such are some of the evils produced by solitary imprudence—evils which have attracted the attention of some of the most celebrated physicians of antiquity, as Hippocrates, Coden, Celsus, which I have already quoted at length, Areteus, Atius, &c. Other physicians, who have enjoyed great celebrity in modern times, confirm the observations of the ancients, and increase the frightful catalogue of ills which the medical writers of antiquity had described. Men of such celebrity as Sanctorious, Lommius, Hoffman, Boerhave, Van Swieton, Kloehof, Mechel, Huller, and Harvey, all of whom have described in vivid and fearful colors the diseases of those who are addicted to solitary vices, must convince the most skeptical. Hoffeland, speaking of young girls who are the victims of this fearful vice, says—"She is a withered rose, a tree whose bloom is dried up; she is a walking spectre."

"How many persons," exclaims the venerable Portal, a physician who published 'Observations on Pulmonary Consumption,' "have been the victims of their unhappy passions? Medical men every day meet with those who, by this means, are rendered idiotic, or so enervated, both in body and mind, that they drag out a miserable existence; others perish with marasmus, and too many die of a real pulmonary consumption."

We have the most indubitably awful proof under our own eyes every week, by reading the deaths reported in the different cities. Notice the preponderance of deaths of those over ten years, from consumption, apoplexy, marasmus, and epilepsy, and you must be convinced there is some unknown secret cause, to hurl so many into eternity every day, from these diseases.

In a work upon the terrible malady of Rickets, the writer says, while speaking on a particular form, that young persons who yield themselves up to the seduction of solitary pleasures are often the subjects of the disease, and he cites a number of cases which came under his own observation. Among them, he mentions that of a young female of seventeen who died suddenly of them, and which were brought on by masturbation. These are his words—"I saw a young girl of seventeen, of puny stature, but who became so rapidly curved, that in six months she was quite hump-backed. The chest was thrust in at the base of the breast-bone; there was a complete hollow at the region of the stomach, while the belly protruded."

A justly celebrated physician of Lyons, who gave himself up for many years to deeds of benevolence and humanity, the sensible author of a work entitled 'Medium of the Heart,' M. A. Petit, seeing the number of maladies which the indulgence of solitary habits produced, thought it was necessary to erect a monument to one of the victims of this vice, not only to avert its danger, but to attest its power in the production of pulmonary consumption. In his preface he says—"Let it be known that pulmonary consumption, whose horrible ravages in Europe ought to give the alarm to all governments has drawn from this very source its fatal activity." And then, in the most eloquent verses, he speaks of the last moments of an unfortunate victim whom he himself witnessed, and whom the tomb of Mont Cindre had closed upon, while yet in the very spring of life. This example is one which has occurred in our own times, and the tomb is placed on one of the fertile mountains which border the Saone on the north-west, in approaching Lyons, and the unfortunate victim who has there found her last resting-place is not yet, in all probability, become part of that dust which has received her. Her miserable error had doubtless been of too long duration; the blow her constitution had received, was too deep.

Baron Boyer, in his 'Treatise on Surgical Diseases,' believed that this constitutional injury may even be prolonged to old age, when this kind of abuse allows its victim to attain it; and that is a secondary cause of many of those cases of dry gangrene, which are observed at that period of life.

These solitary habits in many females produce a swelling of the neck, from the force and frequency of those convulsions which so often follow the repetition of this imprudent act, as well as by the arrest of blood which it occasions in the principal vessels of the neck, in the same way as is observed in epileptic patients. The complexion assumes a yellow tint in some, while others find their skin become covered with scurf.

Professor Richard reports in his 'Chirurgical Nosography,' a very remarkable example of the power of this cause in the production of eruptions:—"A lady had at the same time this pernicious habit and an eruption of blotches. She was advised to discontinue the practice; she did so, and they disappeared. She again took up the habit; the eruption again made its appearance" her reason again taught her the error of her ways, and she once more conquered the penchant, and she was never again troubled with those blotches which had so disfigured her."

Some persons are troubled with cramp in the stomach and pains in the back; some will have pains in the loins and kidneys; others suffer from pains in the upper part of the nose, in the summit and back part of the head, in the groins, as well as in all the limbs; lucorrhœa or whites, acrid and irritating discharges of different kinds, fluxes, hemorrhage of the womb. While some are afflicted for the remainder of their life by relaxation and fall of this organ—pains, at first vague and undefined, then fixed—sometimes dull, at other times excruciating—are in others but the signs of scirrhus or cancer of the womb itself. The belly becomes enlarged, hard and distended; the eyes are surrounded with a leaden-hued circle; the enamel of the teeth assumes a grayish white color, and no longer presents that exquisite polish nor that ivory tint which has caused them to be compared to pearls encased in roses.

A number of painful ulcers are sometimes found on the tongue and the interior of the mouth. A lady had abandoned herself to all the intoxicating enjoyment of solitariness. When she gave way to these excesses her mouth was filled with ulcers of the most distressing kind; when she ceased from these imprudent acts, these ulcers disappeared. The flesh loses its solidity, and becomes flaccid, paleness ensues, wasting wrinkles, inaptitude for all kinds of work or exercise, take the place of the freshness, the soundness, the grace and activity of the body; the bosom, which by its exquisitely developed beauty, shows that the age of puberty and love has arrived—the bosom, whose fullness in the young mother shows that it incloses an abundance of that nourishment so necessary to the tender state of man's infancy; in those who yield to this habit, exhibits nothing but the meagre outline of what it should be, and cries aloud the truth—speaks of nothing but eternity. If such persons enjoy health, they must lose it; if they are attacked with illness, their restoration to health is difficult; if they are so fortunate as to recover they are ever liable to be again assailed. Proper habits are no less necessary for a perfect restoration to health than proper sleep, exercise, and pure air.

Solitary vicious habits have great influence in developing scrofulous diseases. The white swellings which appear in various parts of the body, and filled with a white humor, inclosed in a membraneous bag, formed from the muscular tunicles, are very common: and we cannot doubt that this disorder, which is a species of scrofula, owes its increase to the misfortunes that are inseparable from war and its attendant privations. But what ravages has it not made among persons who have destroyed their constitution by the deadly incentives of a vicious solitude! The hospitals in Europe teem with subjects worn out by the suppuration of these tumors, who owe their frightful fate to this habitual vice. The degeneracy of morals, the absence of principle, and the contagion of example, united with unavoidable privations, have multiplied the forms of scrofulous swellings, which were very rare, and of which the cure is attended with the greatest difficulty, in proportion as these combining causes continue.

Human nature will always be human nature; and necessarily will be mixed up with both vice and virtue. Was not this vice common among the Jews since it so forcibly drew down the attention of their Legislator, who felt himself compelled, in order to restrain its progress, to bring home to the imagination of his people the terrible example of Onan, mentioned in the first part of this work?

The celebrated Joubert, chancellor of the University of Montpelier, one hundred and fifty years ago, complained also, in his Treatise on Popular Errors, of the calamities that this very vice entailed on its votaries of both sexes. Finally, is it not true in this nineteenth century that many dissertations are published on this same subject?

The vicious (solitary) habits then are, beyond a doubt, unhappily too common; but who can say, they were less so in by-gone times! But, however that may be, there is no abatement in their effect. For if we except poisons and some few frightful maladies, the human system has no greater cause of destruction to dread than their sinful secret vices.

Sydenham says, "The organs of respiration are the weakest of all those belonging to the human race; two-thirds of mankind die of diseases of the lungs; and the most common period in which young persons resort to these vicious habits is precisely that wherein the chest exhibits the greatest susceptibility. There is, moreover, a species of consumption to which women are greatly exposed by the very nature of their constitution, such as tuberculose and lymphatic consumption."

CHAPTER XX.

LIFE, WHAT IS IT?—IT IS MERE EXISTENCE WITHOUT HEALTH.

Poets have described life to be the dream of a shadow—a species of flight; have feigned that the gods did not make man a present of life, but that they rather sold it to him. Our life may be truly said to be the dream of a shadow; but this shadow has feeling—is endowed with understanding, and by acting wisely, the dream may be rendered a long and happy one.

That our existence is fugitive, the rapid succession of years, the flight from infancy to youth, from youth to old age, and from old age to the dust that follows it, abundantly testify. But we are in possession of a means of giving to this varying existence—which has so manifest a tendency to its own destruction—a more fixed character; and that is moderation, the source of all virtue and of all happiness. The Author of our being, in granting us life, appears indeed to have surrounded it with numerous duties; but duties are not burthensome, when those by whom they are imposed furnish their tributaries, at the same time an inexhaustible treasure, by which they may be freed from them. With this treasure we are well acquainted; it is deposited within us; we are allowed free access to it, and indeed, are enjoined to draw from it; and if mankind would only make use of it, if they would boldly render themselves subservient to their reason, they would soon discover that nature was really desirous that they should be happy.

It must nevertheless be admitted, that if Heaven has granted to man the splendid gift of reason, it has at the same time rendered it very necessary for him. No other animated being is subject to such numerous diseases, none die from such variety of causes, and none bear within them so many germs of evil as himself. If he knows not how to use those means which have been bestowed upon him for preserving himself from them.

But if we may look upon life as the dream of a shadow—if it be no more than a fugitive existence—if we may in some degree say, not that the Supreme Being has sold life to us—for that would be impious, if taken literally—but it would still seem that it may be so said of those, and that dearly sold too, who have given themselves up as victims of a solitary vice.

Independently of the disposition of the human species towards so many different affections, each individual has a particular tendency to one or more kinds of destruction, by the unequal division of strength, whereby the organs of which they are endowed; and the truth of this is daily attested in our intercourse with the world, when we hear that such and such a one has a weak stomach, a delicate chest, &c.; and as the weakest organs are the first to suffer from the influence of the causes of disease, and as there can be none more powerful than evil practices, it may be safely predicted of all persons yielding to the delirium of solitary vice, the kind of malady their imprudence will bring upon them. In vain it will be believed, that we do not bear within ourselves the "beau ideal" of human organization; and even though so extreme a favor might have been granted to some individuals, it, nevertheless, cannot be contended that such a perfect organization is unalterable. Even Thetis herself could not render Achilles invulnerable throughout the whole of his body.

Bickat, in his General Anatomy, advances a similar opinion, and which has now for a long time been established. He says, "Life is a great exercise, which keeps by desires the various organs in motion, and which leads at length to their repose—this repose is death. But each organ arrives there more or less soon, according to the desire of strength with which it is invested—proportioned to the greater or less disposition to exhaust it—in the course of this great exercise."

CHAPTER XXI.

MASTURBATION PRODUCES EFFECTS UPON THE FEMALE; EASILY DISCERNED BY THE GENERALITY OF PEOPLE.

Persons who are devoted to secret vicious habits, whether they be characterized by the delicacy or the too great activity of the nerves, may rest assured that they will become epileptic, subject to fits, to palpitations, and all other nervous affections.

Any one who may observe a woman who has not been irreproachable in such secret practices, of an apathetic character, weak and languid appearance and pale complexion, you may be assured that she will become idiotic. Mankind cannot imagine the effect these practices have on the present generation, or the destructive influences and consequences that will ensue to those in future.

If pale and sickly children could only be preserved in their early innocence—if those whose parents have fallen victims to pulmonary complaints, could only be kept chaste—if persons who are subject to consumption, would only lead a life of purity, we should soon discover that such disorders of the chest are neither so hereditary nor infectious as they are supposed to be. If the innocent offspring that I have cured of venereal affections, knew the cause of their infirmities and sufferings—their maimings through their whole lives—they would curse —(and I think properly so, too)—the very beings—the very parents—who, as instruments in the hands of Divine Providence, gave them their existence.

My extensive experience, in treating all classes, ages and sexes, for venereal affections, has not only convinced me, but has proven, that a large majority of the scrofulous diseases have been directly produced by the parents or ancestors having had venereal diseases, and did not—though they may have supposed so—have the poison entirely eradicated from the system. All persons, then, who have ever had any disease of a venereal character, should know—by applying to me—whether anything of the kind is still lurking in their system, awaiting an opportunity—from some peculiarly exciting cause—to show its hydra head. All judicious persons will certainly not lose any time in attending to it.

Many of the offspring of venereal affected parents, die in infancy; many are dead when born; but those who do linger, and drag out a short and painful existence, merely postpone their existence a little longer, only to descend at a later period to an early and disgusting death.

I will hasten from this painful subject to one more in consonance with chastity, and refinement of the feelings nature intended us to be possessed of.

Every animated thing—the plant, the brute, the human species—are born delicate and fragile, and grow up in strength; the tender shoot of the oak becomes a hardy tree; the weakest infant—if its source is healthy—a soldier, a laborer, a vigorous man—if nothing be done to prevent it. If nature sleeps, languishes, or is deceived, it knows when to awaken and retrieve each error—if it be not incessantly thwarted or crushed—no matter by what means. The Author of nature has traced out its steps, has dictated its laws; and a premature death, the melancholy consequence of evil practices, is only one of the rules of his immortal code; let them, therefore, be engraved on the heart.

How much it were to be desired, that persons would not so frequently despise themselves in the world, and that they would not invariably attribute to family disease, or the contagious nature of consumptions, that which is traceable to vicious habits alone. Persons born of healthy parents, of sound constitutions, and who themselves are quite robust, and always breathing a fine air, find the chest affected by their evil habits, and the grave opening before them, long before their time; while, on the other hand, they on whom heaven seems only to have bestowed a very short life, are preserved, by a purity of conduct, to the limits of extreme old age.

Far then from asserting that persons afflicted, as previously mentioned, are to be despised or blamed; I think, on the contrary, that with some (as I have there stated,) it is more the result of an unhappy destiny than merited; and that every motive of benevolence urges to pity, and a relief of their condition. The blame we may feel disposed to attach to some persons for their bad formation. You will remember, I have mentioned that rickets are caused by masturbation—can only be addressed with justice, to those whose childhood and youth have not always been free from reproach. It is easy to distinguish between those whom we ought only to pity, and those whom we ought, at the same time, to blame. The former have the head large, relatively to the rest of the body, and the arms long—as it were, drawn out—the latter do not show these peculiarities. They first are usually gay and lively, while the second, on the contrary, combine with their deformity extreme stupidity and inertness. All functions, except respiration, are well discharged by the former; with the others, all are accomplished badly and languidly. These always appear to be laboring under a severe malady, and are incapable of any toil; while those are wholly free from this appearance, and daily engage in occupations more or less difficult—in study, in the cultivation of science, and the practice of the arts.

What will become of a lady's beauty when her health is deeply impaired? It can only bloom and disappear like the withered form and faded color of a flower nipped in the bud; or, to make use of another comparison, like the ruins of a temple destroyed by profane hands. It loses the elegance and majesty which delighted the imagination and inspired respect long before years and decay would have impaired those qualities. A young woman is this temple, and must expect to lose all the attributes of beauty, when once she gives herself up to this most destructive of all passions. The growth of the body, the development of the figure, all grace and freshness, will disappear; for this error spares no charms. No doubt the words of a celebrated physician, respecting a votary of this solitary vice, will recur to every mind—

PLATE 25.

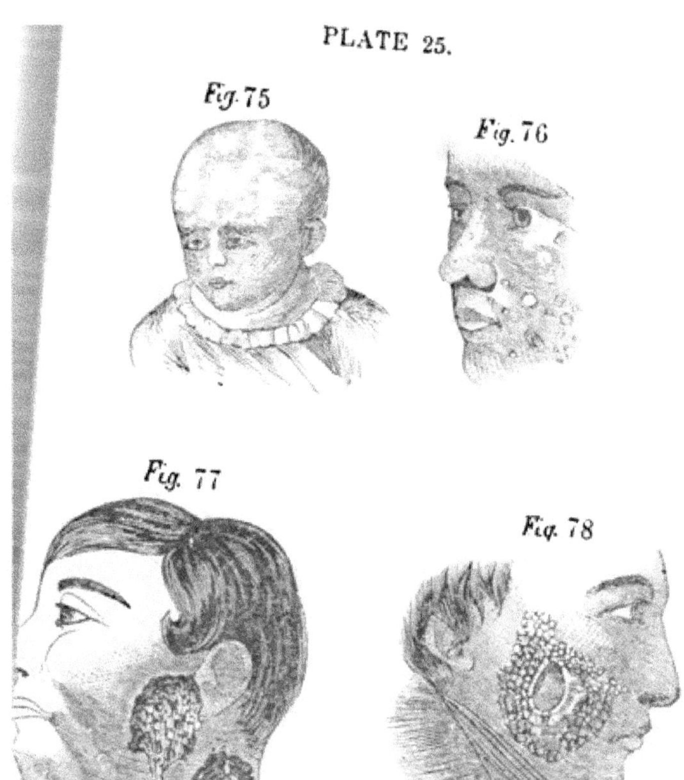

"She is a tree withered while blossoming—a perfect walking skeleton."

What a different picture those young females present who display so agreeably all the charms of their tender and admired sex. Consider a girl at the age when the attractions of youth succeed the grace of infancy; happy is the maiden who preserves her primitive purity, when, like an unknown lake, in the bosom of a lovely country, her imagination as yet has only reflected the heaven above and the verdure and the flowers around. She appears to unite in her person the rarest charms of all the universe.

If man be the Lord of creation, such a female is surely the queen. But this grace, and this fresh and fair complexion, are never to be met with in persons consumed by disease; and these portraits, presented to us, under the name of goddesses, are those of persons who have really existed. Thus Fontaine, when he covered with roses and lilies the Venus whom Homer, three thousand years before, had begirt with the enchanting zone, was no doubt thinking of the young and virtuous La Sablière. And probably Virgil was indebted to some princess of the court of Augustus for the majestic air which excites our admiration in his Dido.

Can health and grace exist without innocence and modesty? Chastity is the daughter of Modesty, and therefore health cannot long be enjoyed without chastity. Socrates said long ago, that a handsome body gave promise of a noble soul. Some ancient philosophers, have supposed the soul to be a kind of divinity inclosed in our bodies; certainly the sweetest rays of the Divinity are found nowhere so conspicuously as on the brow of a woman worthy the esteem and confidence of her husband. One of the wisest men says, "The beautiful of every description is beautiful of itself. It exists in and of itself; praise forms no part of it. Thus nothing becomes better or worse from the opinions of others."

CHAPTER XXII.

CAUSES OF FEMALE PROSTITUTION.

The same remarks that the reader has found in the previous pages of this work, relating to the false modesty heretofore existing on the subject of masturbation, and which has stifled the horrid truth relating to it, from sooner having become known to the world at large, is applicable in the fullest sense to the awful subject now before us

It is, like masturbation, one of those subjects so fenced and hedged around by ignorance and false modesty, that the very silence and darkness allowed to pervade and surround the disease, has continued to propagate and augment its ravages. The least familiarity with the customs and habits of the city, or a knowledge of a tithe of the haunts of vice, will reveal the fact, that tens of thousands are living amongst us by open and private prostitution. What might be considered as gross satire and superfined didactic criticism, might be translated and published, and the community could not deny that it was a fair description of every-day life, as it exists in the pseudo moral atmosphere of almost every city and country.

The very virtuous and ignorant may occasionally hear of some shocking case evolved in the horrid dens which abound in the large cities, or seeing some of their bloated and disgusting inmates, may be thankful, with complacency, that the vice is confined to those localities, and to such self-degraded objects; and as a writer in a daily journal said, "If victims be necessary for such a *Moloch*, would that such were the case."

While such a horrid evil exists, is it not surprising that no remedy has been found and applied as a cure and preventive? Those females, who, in their conventions advocate the rights and claims of their sex, with an eloquence equal to the efforts of the sterner sex in their political struggles, have entirely overlooked this subject, yet it more vitally concerns their honor and the welfare of humanity.

Mr. Fowler has battled, and done well, with this monster, whose victims are innumerable, and I learn has been ably assisted by Mr. Wells, Mr. Woodbridge and Mrs. Gove, with a degree of womanly courage and modesty, which cannot be too highly praised. I have no doubt others would have joined in the noble work, but for the erroneous idea, the more publicity you give to the subject, the more would be its increase and dissemination. It is to be hoped, however, that this argument will be forever refuted, before many more victims are sacrificed from its belief, for I am pained to say the range of this could hardly be extended any further than it is at the present time.

There is too much sickly sentimentality afloat, in regard to the unfortunate females who prowl about the streets of cities in search of their prey. There are undoubtedly among them those who are the victims of false promises, and of vile, heartless, and accursed seduction; but the number, I think, is comparatively small. It is, in fact, very seldom that men or women descend at one step from virtue to the lowest condition in vice. Persons in the height of passion, will slay their oppressors or adversaries; but the mind requires to be familiarized with guilt, and the walls of virtue require to be sapped and undermined, before they tumble to pieces. Men neither become drunkards, robbers, forgers, nor counterfeiters, all of a sudden. They have trained and habituated their minds to crime before its commission. Their fall may appear rapid, but if the truth were known, it has been by a road that they had labored hard to prepare. It is just the same with female prostitution. The idea of a female that is virtuous, at once betaking herself to the streets, is preposterous. This, therefore, leads us to consider the way by which females reach this sad state of abasement.

Speaking of early vice, Mr. Fowler says, "I would not defame my race, but facts extort the reluctant declaration, that few have more than the faintest conception of the fearful extent to which this vice (masturbation) in all its appalling forms is practiced. It is the destroyer of our youth of both sexes, and still more of our husbands and wives." Chatecise, promiscuously, every boy you meet, and then say, if nine out of every ten, from *eight* years old and upwards, do not practice it more or less? and I have not the least doubt that nearly every one does so, after they have arrived at the age of puberty. No child is safe from this loathsome habit; and, as I have previously shown, our schools are especially the nurseries of this vice.

Mr. Woodbridge, in the 'Annals of Education,' says, "The fatal vice is spreading desolation throughout our schools and families, unnoticed and unknown." Our boarding and day-schools are sources of untold mischief.

A writer says, that "at West Point, the mental debility occasioned by this vice was the reason why so many of its students were unable to pass examination." "But," continues Mr. Fowler, "our families at least are safe. Exclaims the fond mother, 'My daughter's native modesty is her shield of protection.' Would to God this were so; but facts wrest even this consolation from us. They may be less infected; yet women, young and modest, are dying by thousands, of consumption, of female complaints, of nervous or spinal affections, of general debility, and of other ostensible complaints innumerable, and some of insanity, caused by this practice."

Mrs. Gove, in her 'Lectures to Ladies on Anatomy and Physiology,' says—"About eight years since my mind was awakened to examine this subject, by the perusal of a medical work that described the effects of this vice when practiced by females. This was the first intimation I had that the vice existed among our sex. Since that time I have had much evidence that it is fearfully common among them. There is reason to believe, that in nine cases out of ten, those unhappy females who are tenants of houses of ill-fame, have been victims of this vice in the first place. Professed Christians are among its victims."

Here, then, is the fountain from which prostitution flows. Need we wonder at the number of its votaries, or be surprised at the unhappy marriages of which we daily hear, and at the frightful mortality among children in the early months of their existence? How do parents discharge their responsibility in guarding, by caution and advice, their children, from this deplorable vice and its hideous consequences? In the familes of Quakers, and in strict Catholic families where the duties of the confessional are attended to, the uses and abuses of the instincts, passions and affections are pointed out to children, and they are taught to regulate and control their desires and appetites. In most families, however, children are left unguarded against the bad example of their elder associates, and habits are acquired and formed, in innocence and in ignorance of many evil consequences, either moral or physical.

It would seem sufficient to inform children of the baneful effects of a vice which exhausts the body, destroys the eye-sight, impairs digestion and circulation, deranges the brain and nervous system in an astonishingly short time, thereby impairing the mind, destroys the possibility of a healthy offspring, and stamps the face with its marks and signs as visibly as does the small-pox. Let parents and guardians exert themselves to avert this wide-spread and insidious contagion, and use their influence to promote marriage, and our streets will cease to be filled by wanton harlots and licentious profligates.

My own practice is constantly presenting me with the most heart.sickening cases of females who were once of the highest respectability. The last case of this kind was very recent.

CASE LIV.

It was that of a young female.teacher residing in this city. Her ardent temperament was the cause of her commencing the practice, which led on to the greater vice of unchastity, but still practicing masturbation at frequent intervals, till, when she applied to me, she was a mere wreck, of what, she said, she once was. Her case presented the symptoms which are portrayed in so many of those cases, in the previous pages, involving the whole system. To effect a cure for her, required patience and perseverance.

If the size of this little work would permit it, I would detail a number of the female cases from my note-book, as they are there entered in full, but as it will not, I shall content myself with presenting the case of a young lady, which it pains me to say, was beyond the reach of human ability to restore, or even to prolong her existence for any very great length of time.

CASE LV.

About a year ago, a lady called at my office, and after inquiring if I was the doctor, and if so, whether I was alone, that she might have an opportunity of privately presenting the case of her sister to me, for my consideration and advice. After requesting her to be seated on the sofa, and showing a patient out by a side door, I heard her statement, which I here copy in as few words as possible.

She said, her sister was about thirty years of age, and had been sick a long time; but what the real difficulty was, her folks could not find out, neither from the physicians who her father had employed, nor from her own statement. Her parents were in wealthy circumstances, and had spared no pains or expense in their endeavors for her cure. As her condition had grown worse, while under the charge of their regular family physician, they called in others for consultation, until they had employed the most eminent practitioners of the present age. She had the symptoms of almost every chronic disease, which they of course endeavored to remove. The cause which produced all this disorganization, however, they were ignorant of, and therefore could not remove the effects.

After continuing in this miserable condition for a number of years, she attempted suicide by taking laudanum, but her attendants made the discovery in time to frustrate her designs, by the physicians applying the stomach pump, and using counteracting remedies. In a few days after this, she disclosed the real cause of all her troubles to this sister, on condition that she would not reveal it to her parents. Three of the most eminent of her physicians were then made acquainted with the secret, but their success was no better than before. She nearly extinguished her feeble lamp of life, after she found they could not be of any benefit to her, by a second over-dose of laudanum, but which was frustrated in the same manner as before Her nerves and mind had been so weak for a long time, that they had prohibited her entirely from reading newspapers. One day, by accident, she got one in her possession, which contained my advertisement; seeing which, she without losing a moment, sent her sister to me.

If her parents had only known of her complaint in time, so that my services could have been engaged early enough to have been of some avail, they might have been saved the loss of their daughter, the heartrending scenes and trials, and a very large expense Let no person, then, after reading this little unpretending volume, allow a single day to pass, without procuring for themselves, their children, or friends, that professional advice which enables me to save thousands from an untimely grave.

At the time the sister applied to me in her behalf, she was wasted to a mere skeleton. Her stomach was so weak that she could retain nothing but liquids in it. The sight of a physician would bring on spasms. She could only be restrained by force from practicing her loathsome abuse. Her mind was extremely idiotic; she was, in fact, insane.

This being her situation, her case was hopeless, as she could not endure or sustain any treatment that could be of any permanent benefit, though her father offered any fee for her restoration.

CASE LVI.

This female patient and a man, thirty-eight years of age, who came to me in just about the same state, something over two years ago, are the only patients I have ever had which I could not cure; and I have had many of the most deplorable patients whose disease arose from self-abuse, and an innumerable number of horrible and loathsome cases of syphilis, which had proceeded so far as to cause the destruction of the bones of the limbs and face.

Dr. Larmont:

Dear Sir,—It affords me no small degree of pleasure to cheerfully bear testimony to your high private character and standing; and in a professional point of view

I am proud to acknowledge your superior knowledge and skill in the treatment of gonorrhœal, syphilitic and spermatorrhœal diseases, as well as the various complaints of females. In fact, sir, it is due to you, that I should speak what I know to be true; that you are in this branch of the medical profession in advance not only of the profession, but also in advance of your worthy and illustrious preceptors, the great Ricord of Paris, and the great Acton of London.

I say this not to flatter, but because I have frequently seen the happy results of your superior treatment in the worst, most unpromising, and cases cast off and regarded as incurable by able surgeons, in this and other cities of the United States, and in Europe.

Yours,

Most respectfully,

H. A. Smith, M. D.

As I have previously stated, the originals of the above and other testimonials, can be seen at my office; or, those residing in the city can be seen personally, if required.

Dr. Larmont:

Dear Sir,—Your very extraordinary success in gonorrhœa first led me to cultivate your acquaintance and friendship. Nor do I regard your present disinclination to make your plan of treatment publicly known to the profession as ungenerous; for what you have toiled to acquire is yours, unquestionably, and of right.

You say in your last letter that you hardly know whether to edit Ricord's great work on Venereal—hereafter, by adding copious notes, or give the profession one entirely your own.

Now I regard all or nearly all the works since the publication of Ricord's book, as copies, compilations, or senseless alterations of the above, Acton and a few other valuable works.

In fact, doctor, we have all of us secrets in practice, which we want to reap benefit from before allowing others to enrich themselves by them; yet I am happy to learn your inclination to do the profession so great a benefit, and through them suffering humanity an inestimable kindness, by by-and-by publishing your vast experience in treating diseases of the genito-urinary organs.

I must again acknowledge my indebtedness for the valuable hints on gonorrhœa and its abortive treatment, which I have found to give perfect relief in a few hours.

I am now about to go to Dublin to remain a few days, and when I return to London I will again write.

Very respectfully,

J. CURTIS, F.R.S.

P. S.—I agree with you in denouncing unnecessary surgical operations, and particularly the early opening of buboes, which your plan precludes the necessity of doing. J. C.

DR. LARMONT:

My Dear Doctor,—Yours of Feb. 21st I received by our Havre steamer. You seem to regard unfavorably the plan of cauterization in urethral strictures I too must confess much disappointment in the indiscriminate use of caustic, which I was early led into by its popularity in Paris and Lyons. I think if Lallemand had lived, that he would have advanced very much the unsuccessful treatment of Spermatorrhœa, which he brought out of obscurity. Your mode of cure seems, in such cases, very novel, and peculiarly your own. I will give it a trial, as it looks plausible.

So far as chancre is concerned I unhesitatingly say you bear off the palm of victory in treating chancres. We, on the continent, cure as surely as you, perhaps but by no means so soon.

Some of my brethren were skeptical in reading your cases, but Dr Hill assures me you are doing wonders

Au Revoir,

VIRTUE AMI

CHAPTER XXIII.

Certificates given me by the patients themselves, for curing them of Emissions, Seminal Weakness, Impotency, and General Debility, after they had been under the charge of a number of Physicians, and used all of the quack remedies, such as Cordials, Elixirs, Antidotes, Drops, Compressions, and Mechanical Instruments.

CASE LVII.

——, Conn., May 19th, 1852.

Doctor Larmont cured me of Emissions, and a Stricture of the Urethra, in a few days, by local treatment only.

I commenced the practice of masturbation when about twelve years of age, and continued it till within a few months. I am a member of the church of Christ, and did not suppose I was doing anything more than to satisfy, in an innocent way, the promptings of an ardent temperament.

The disease had become so general as to cause indigestion, extreme nervous debility, irregularity of the bowels, swelling of the testicles, pain in passing urine, and a disagreeably painful sensation of the bladder. My mind was affected to that extent, as to cause a partial loss of memory. Before calling on you, I had taken the advice of a physician in Rochester, New-York, and two others in Boston, Mass., neither of whom could tell me what was my real trouble. I forgot to say, my age now is twenty-four. WM. CORNELL.

CASE LVIII

——, Conn., June 7th, 1852.

M. LARMONT:

Dear Sir,—Yours of the 4th came duly to hand. In reply I would say, I have had no emissions since I saw you. Yours, respectfully,

J. D.

I will only say, in reference to this case, that he is

about twenty-six years of age, and commenced the practice of masturbation very young. Previous to his coming to me, he had been under the charge of three other physicians for a long time, notwithstanding which his emissions occurred every week. He had attempted connection with a female a number of times, but the semen was ejected so soon, that he never could accomplish the act till I had cured him of the emissions. He put himself under my treatment in March, having in compeication, indigestion, costiveness, a breaking out on his face, and nervousness.

CASE LIX.

U——, New York, June 20th, 1852.

SIR,—I write and let you know about myself. I have not had any more of the emissions since I wrote you last

Yours, M.

M. LARMONT, Esq.

This gentleman came to me in October, 1851. In January, 1852 he wrote me he had within that time had two emissions: his next letter is the one above given. He was about twenty-eight years old, and had, previous to coming to me, been so affected generally, as to cause indigestion, costiveness, great debility, affected memory, and a hanging down of the testicles.

CASE LX.

DR. LARMONT:

My emissions have been stopped by you in two or three weeks, after having been under the treatment of others for years. My health was never better.

Thankfully and confidently yours,

J S——D

New York, Jan. 15th, 1851.

This patient had been married a little over a year before coming to me, but was so impotent that he had up to this time been unable to cohabit with his wife, although his general health was not much affected. He was 23 years old when married, and was advised to it by his physicians as being the only cure for him. Well, when he could not cohabit, they advised him to go south a year for his health, and he did so; and when he returned was no better. He then came to me, and the above certificate tells when he was able to perform his family duties. He would be happy to verify the above to any one that will call upon him personally.

CASE LXI.

Milwaukie, Wis., Aug. 8th, 1850.

DR. LARMONT:

Dear Sir,—Having arrived in Millwaukie I with pleasure fulfill my promise of writing to you.

I am happy to inform you I have received great benefit from your medicine, and am in a much better state of health. I am, sir, most respectfully,

Yours, &c.,

J. E D.

This young man applied to me in July, about four months after the above letter, he wrote me again, saying he had had no emission, although he had lost a part of his medicine, and wanted me to send more, if I thought it necessary. I did not send anything, but wrote him to let me know if all was not right. I have not since heard from him. He had been treated in Europe before coming to me. His age was twenty.

CASE LXII.

O——, New York, June 18th, 1851.

M. LARMONT:

Dear Sir,—I feel it my duty to inform you of my health since I saw you. I have had but one emission, and that was a few weeks after I got home.

I am, most respectfully,

Yours, —— Y——.

This young man was about the same age as the next preceding one, and was much troubled with priapism, weakness of the stomach, deranged bowels, and those tell-tale pimples on the face. I had a letter from him last fall that he was entirely well.

CASE LXIII.

L——, Indiana, Oct. 6, 1850.

Dr. Larmont:

Sir,—I received yours in due time, and commenced using your pills as you prescribed, just four weeks since, and they have had the desired effect, as far as night emissions are concerned; but there is yet a looseness of the bowels. Just before getting them I had a spell of sickness—billious diarchœa; but I am now in perfect health, and getting stronger.

I remain yours, sincerely, C——.

This patient had been afflicted with a chronic diarrhœa more or less for a year; he was emaciated, and without a doubt, a rapid consumption would have soon carried him out of this world if he had not seen my advertisement, and applied to me at the time he did. I had a letter from him a few months after the above, saying, he was married, and in good health. His age was thirty-three.

CASE LXIV.

Dr. Larmont cured me of a long-standing seminal weakness, and emissions, complicated with stricture, and that too in a fortnight, so that I had but one emission after commencing his treatment. I had previously been under treatment to no benefit.

C. J. C——, Brooklyn.

July 1st, 1851.

This patient was nineteen years old, and of a scrofulous constitution.

CASE LXV.

Dr. Larmont:

You cured me of gonorrhœa of some time duration, by a single local application, which was complicated with a bad stricture. By my early improper habits I was afflicted with emissions and great debility, which were both also cured by you in a short time. I had been for a long time under previous treatment to no benefit. F—— R——.

New York, June 23d, 1851.

This young man was twenty-three years old, and belonged to Nova Scotia, but came on here for medical advice, and has procured a situation in one of the largest establishments in this city, where he can be seen by the afflicted at any time.

CASE LXVI.

For over nine years, I was troubled with seminal emissions, which caused a general derangement of my system, such as nervousness, affected memory, costiveness, depression of spirits, &c. Within the above time I used all the advertised remedies for these complaints, and was under the care of two other doctors, all to no benefit, till I came to Dr. Larmont, who has cured me entirely, having had only one emission since coming under his charge. My health is now as good as it ever was, yet it is only two months since I was cured.

A—— D——.

New York, Sept. 23d, 1851.

The above patient was troubled with a constant discharge of semen and mucus, arising from an inflammation along the whole canal of the penis, and particularly in the region of the seminal orifices and neck of the bladder. This complaint often involves these parts and the prostate gland, not only seriously, but so much so as to endanger life; therefore, not only is the correct treatment necessary, but it must in many cases be prompt and decisive. All those interested will hereafter be able to understand the danger of cordials, elixirs, mechanical instruments, and all quack remedies.

CASE LXVII.

New York, March 3d, 1851.

Dr. Larmont:

You have cured me in a few weeks of great debility and nocturnal emissions that I was troubled with for years. I have had but one emission since being under your care. I was under treatment a long time before calling on you, and that without benefit.

W—— H——.

This man was a journeyman tobacconist, and chewed as many as six papers of tobacco within the twenty-four hours, and, in addition to that, smoked a number of cigars. I was obliged to gradually break him of this destructive habit, as his nervous system had been brought by this enormous use of tobacco, the weekly emissions at night, and the escape of semen while at stool—produced by the slightest straining—to the most deplorable condition imaginable

CASE LXVIII.

I hereby certify that Dr. Larmont, 42 Reade street, has cured me of three bad strictures, which, with the gradual wasting away of the semen, and emissions at night, caused impotency and very great weakness of my whole body, costiveness, nervousness, impaired memory and depressed spirits—all of which he has cured me of in a few weeks without my having had one emission, after commencing his treatment, which was without any medicine. P—— N——.

New York, Nov. 25th, 1851.

The above case was an extremely dangerous one, yet, it will be seen, my treatment acted upon him almost like a miracle. I often astonish myself at the rapid cures, and especially when I am not obliged to use a particle of medicine in many of this kind of cases, not even to regulate the bowels, for they seem to correct themselves immediately.

CASE LXIX

Dr. Larmont:

You have cured me in a few weeks of an old seminal weakness disease that other physicians had treated, without benefit, for a long time. I was very weak, and had emissions very often, but since your treatment, have had only one. My health is better than it has before been for years. F—— G——.

New York, March 4th, 1851.

This case was nearly a type of the preceding one, and, as I have never taken a certificate for a cure of any patient who had not been treated by from one to twenty other physicians, I shall copy many of them without comment, for my professional duties are so arduous that my time will not permit me to make the comment I otherwise would and should. My readers will understand by this that all of the certificates I publish are for cures that had been abandoned by others, or considered incurable.

CASE LXX.

Dr. Larmont cured me of stricture, seminal emissions, chancres (syphilis,) and vegetations (warts) in a short time, without scarcely any medicines. I had been under other treatment for some time, to no benefit. I was weak, and my system entirely disordered. I have had but one emission since being under his charge. I have now been cured four months. John M——.

New York, October 21st, 1851.

CASE LXXI.

Chicago, December 19th, '51.

Doctor:

Dear Sir,—Received yours, but not able to write, being under treatment of sore eyes. I think that I am pretty much freed from the emissions. It has been four months or more since I had a discharge or felt like it. I feel much better than I did, and feel much relieved. My flesh don't twitch as much as it did. Once in a long time I still feel something wrong in the pit of my stomach, otherwise I think my complaint is pretty much stopped. I am still careful about my food.

Yours, M—— P——

My readers will discover that the language of the above is quite erroneous, but I prefer to always give the exact language of the patients themselves. His age was thirty, and his case was a very severe one; he had previously been treated, both in this and other countries, for a number of years.

CASE LXXII.

Ohio, December 16th, 1851.

M. Larmont, M. D.:

Dear Sir,—So long a time has elapsed since I have written to you, that I presume you had given up all idea of hearing from me again. I was married, on the day I wrote you was set for it; yes, married. This will surprise you, when you reflect what I have said; but it is, nevertheless, true. I no more expected to be able to perform what is expected of every married man than I expected to swim from New York to Liverpool. I did not, as I supposed I would, have an emission before an entrance was effected, and not, I think, until the proper time; but, notwithstanding your advice to have connection but once a week, I have had connection some five or six times, in less than a week. I have no further doubt about my case. Very respectfully,

H. D—— S——.

The letter previous to this one, was written some six months before. As will be seen by his letter, he was one of those who look on the dark side of everything, and could not be convinced that he was well till he had been married some time. He had great reason to doubt, however, for he had been unmercifully plucked by the regular physicians, as well as the innumerable charlatans abounding in all cities, for a number of years before coming to me. His semen having wasted away for so long a time, left him entirely impotent, before my treatment. He had all the other symptoms accompanying these extreme cases.

CASE LXXIII.

C——, New York, Dec. 1850.

DEAR SIR:

I took your advice and did not put off the marriage-day as I thought I would have to when I saw you at your office, on my return from Philadelphia. After returning home, I applied myself to your directions, and was more surprised at the rapidity of my recovery than you was sanguine of your ability to cure me. I hope you will excuse me, for the fear I expressed that you might be one of those numerous impostors, who assume the name of doctor, and swindle those unfortunate victims who are ignorant of the pretenders and quacks in your city, Philadelphia, Boston and Albany, who swindled me out of a large sum of money, besides leaving me in a worse state than when I commenced their loathsome nostrums. Well, as I said before, the wedding day came and passed, and instead of failing to cohabit as often as you said I might, I have overstepped the bounds very materially, but will pay a more strict observance to your directions hereafter. It is now about four months, and I feel as strong and well as any man; my flesh is returning, so that my weight is increased about thirty pounds. You of course received the balance of your fee, or you would have written me before this time. Yours, thankfully,

R— G. M—.

This gentleman was about thirty-five years old, at the time he placed himself under my charge, and was totally impotent—the semen wasting away by degrees, bowels disordered, troubled with flatulency, a gulping of wind from the stomach, indigestion, a buzzing noise in the head, dizziness, and a faint kind of weakness in the stomach. In about five months' time he had recovered his strength, which he possessed at the time of his writing to me.

I could go on, and fill a volume like this little book, with extracts from the letters of patients that I have never seen even, acknowledging my curing them, the same as those herein copied. Enough of them, together with the certificates also copied, will satisfy any person however long he may have suffered, or the many physicians and remedies they may have employed, that my treatment is not only certain, safe and mild, but probably the only sure mode of cure, at present known. The subject requires me to be plain in my language, and some may, therefore, consider me egotistical, still the truth nevertheless requires it.

As a concluding part of this work I shall copy a few of my certificates of the unparalleled cures performed by me of the other class of private diseases known as *venereal*, so that those who are unfortunately afflicted with these dangerous, hereditary and flesh-destroying poisonous diseases, can have more than the full proof of what I assert I am doing daily.

I overlooked two certificates of very remarkable cases caused by masturbation, and which the gentlemen who received that benefit from me, which saved their lives, and the ruination of their families, wished me to publish with their full names, the better to convince the extremely skeptical. I have published them a great number of times in the newspapers, within the last two years.

CASE LXXIV.

New York, December 5th, 1849.

Dr. Larmont:

Dear Sir,—I must say that I consider you a rare dispenser of health, for, after spending over $300, (three hundred dollars), with the physicians of this city, Boston and Baltimore, for the cure of impotency, emissions and general debility, all the time getting worse under their treatment, I fortunately saw your advertisement; and this is to certify that, after being under your care about a month, I found myself entirely cured and once more a man. I send this in order to let those afflicted know where they can get a speedy and permanent cure.

Yours, truly,

James Salisbury, Mate.

This man had been married about five years, but had no children.

CASE LXXV.

BELOVED DOCTOR: New York, Oct. 1st, 1849

If I was the possessor of a thousand gifts, I would ask you to accept every one of them, besides the small fee which I paid you, as a slight recompense for your invaluable services in curing me of the most deplorable effects of self-abuse, commenced when I was young. The emissions at night, and leakage of the semen, costiveness, indigestion, and, as I said, all the effects of the masturbation, you have cured me of in a few weeks, after I had been under half-a-dozen physicians for some six years, besides using every quack medicine advertised, all of which cost me hundreds of dollars, my little all, besides what I could spare, after supporting my family. The friend you had cured that sent me to you, I cannot thank enough. Yours, forever

JAMES EVANS

CASE LXXVI.

DR. LARMONT:

I was under the care of three physicians for a year, and used everything I could hear of, such as the swindling remedies, so impudently advertised by the pretended doctors, for the cure of syphiltic ulcers on my body and face, which were covered, and very painful. I took the disease by a razor cut, in the hands of a barber, affected with primary syphilis. My whole system was terribly affected, and the urinary organs and lower part of my body pained me constantly. Instead of my water being natural, it was like matter, and followed by a milky discharge. After suffering so long, and still getting worse, I was sent to you, and thank God, that through him you saved me from a lingering, disgusting disease, and an awful death. I am happy to say, you have cured me in a few weeks, so that I now look and feel younger by ten years than I did when affected.

WILLIAM FERGUSON, Perth Amboy, N. J.

New York, Feb., 27th, 1851.

This case speaks volumes in favor of my Paris and London treatment of venereal diseases, as there are no

PLATE 26.

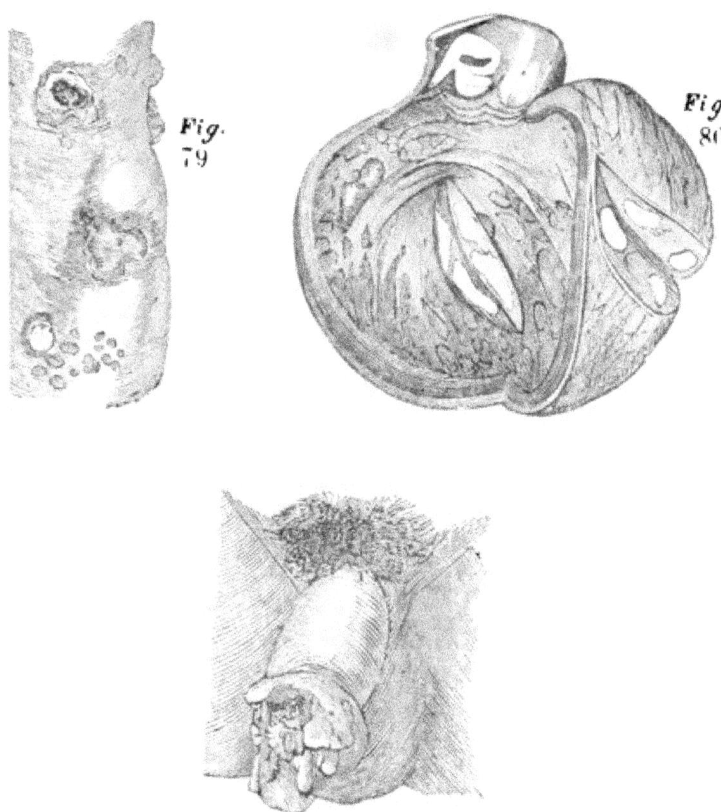

remedies employed by any one else in this country or Europe, that are so rapid in arresting and destroying these poisonous diseases. My remedies do not interfere with the business of any one. And as a proof that I do not use mercury, this man was in the water nearly every day, as he is an oysterman. I often have patients that cannot get off of their beds, from pains and weakness, up and attending to their business in less than a week, for instead of reducing them, I tell them to eat as much as their appetite craves. This man has now been well about two years, and his health is as good as it ever was.

CASE LXXVII.

DR. LARMONT:

Dear Sir,—About one year and a half ago, I contracted a chancre, (pox), for the cure of which I applied to a number of physicians without receiving any benefit. After being under their care for this length of time I applied to you, when I was covered with ulcers. After being under your charge for four weeks, I was entirely restored to health, and have never been better in my life than since then. JOHN ROLLISON, JR.

New York, May 11th, 1850

This certificate was given to me by him, about nine months after the cure, on his coming to me for the cure of a primary affection, he had caught a couple of weeks previous, and which he had had about a week. In a few days he was entirely well of this new infection. His weight at the time he gave me this certificate, was forty pounds more, than when I cured him nine months before.

CASE LXXVIII.

DR. LARMONT:

You cured me of a gonorrhœa of some time standing, complicated with an acute swelled testicle, in a few hours. I have continued my work, and yet the testicle has nearly resumed its natural size in this short time.

P. BROADHEAD.

New York, August 20th, 1850.

The above will satisfy any one that my vegetable application not only brought away his clap in a few hours, but cured him of the most painful affection, swelled testicle, that ever accompanies this disease.

Now, just compare this simple, but only rapidly efficacious treatment, as yet known, with that which is followed—almost universally followed—by the Profession. The first thing they would do for you, would be to restrict your diet to mere nothing, make you gulp down salts and senna, or some other powerful laxative; the next, they would apply from five to twenty-five leeches to the affected part, and if there was a general fever, apply the lance, and rid you of so much blood, that it would take you a long time to recover your strength. Well, after the inflammatory symptoms had subsided, then you would be ready for that, to them, heroic remedy, the stomach-revolter, copaiva, mixed up with cubebs, turpentine, nitre, &c., &c. In the course of three or four weeks the inflammatory symptoms would probably be gone, and the chronic state assumed; then would have to come the injections, such as the nitrate of silver, sugar of lead, zinc, tannin, alum, &c. In the course of another few weeks the discharge probably would be so much reduced as to leave merely a little in the morning, such as a drop, or the lips of the canal glued together at that time, and may be, just before passing urine. Every physician who knows anything of these diseases, will now tell you, there has a stricture formed in the canal, which must be cured, or it will put an end to your existence in a most distressing manner Suppose a person don't care for all these pains, dangers, and loss of time, he would be placed in rather a peculiar situation if he was married, or the wedding-day appointed, surely. Those who have ever had this complaint, and been under their treatment, know this to be a correct statement of thousands of cases.

CASE LXXIX.

Dr. Larmont:

Dear Sir,—This is to certify, that you cured me of a gonorrhœa in one day. I had previously had the disease for some time. D. M. Corbyn.

New York, June 8th, 1850.

CASE LXXX.

This is to certify, that Dr. Larmont cured me of a gonorrhœa, chancres, (syphilis), and a large inflamed bubo, without the bubo being opened. He drove it away entirely in a little while. The gonorrhœa he stopped in a few hours. I had the disease some time before coming to him. CASSIMER DEROUD.

New York, July 20th, 1850.

CASE LXXXI.

This is to certify, that Dr. Larmont cured a gonorrhœa for me, by a local application, after I had been under the treatment of another doctor for some time, continually getting worse. W. N. CRAFT.

New York, March 13th, 1850.

CASE LXXXII.

Dr. Larmont stopped a gonorrhœa for me at two different times, by one local application at each time.

J. F. S——

New York, April 8th, 1852.

CASE LXXXIII.

Dr. Larmont cused me within twenty-four hours, of a gonorrhœa, without pain, by a single local application He done the same about nine months since.

L. L——.

New York, Feb. 18th, 1852.

CASES LXXXIV—V, VI, VII, VIII, and IX.

I first came to Dr. Larmont after I had been under a number of physicians for a weakness of the canal of the penis, caused by a gonorrhœa not being cured soon enough, as I had not been properly cured for some three years. Dr. Larmont then cured me in a few days, and has in five other cases done the same within twenty-four hours, with one local application.

J. T. W——.

New York, July 26th, 1851.

CASE XC.

Dr. Larmont:

You cured me of a gonorrhœa with a single application, and four doses of medicine, in twelve hours. I had before had it for six months, all the time taking medicine. Wm. M——.

New York, Sept. 6th, 1851.

CASE XCI.

This is to certify, that Dr. Larmont, of 42 Reade st., cured me in two weeks of syphilis, after I had been under the charge of Drs. H——, P——, W——, R——, S——, C——, and R——, for more than a year, without getting any better, notwithstanding their giving me mercury, potassia, copavia, &c. J. M——.

New York, March 25th, 1852.

Those afflicted with syphilis, can judge of the imposition and swindling practiced upon the innocent though ignorant persons, by pretenders, assuming the title of physicians, the better to palm off by long advertisements in the papers, their quack nostrums, for this patient, and many others who have given me these different certificates, used the two prominent ones, besides being under a number of family and special practice-doctors. It will be seen, that all combined could only rid him of his money, and not the disease.

CASE XCII.

About two years ago, I called on Dr. ——, alias —— and ——, of D—— street, on the appearance of a gonorrhœa, who agreed to cure me in a few days, but instead of that, after using injections and medicines for a long time, he ended it by leaving me with a gleet and stricture, which he kept on treating me for, till I had been under him for about eight months, and paid him about $20. I left him and went to Dr. —— F—— st., and was under his charge for about three months, and paid him about $30. But, as I was getting worse, I left him, and went to Dr. ——, A—— street, staid under him about six months, and paid him about $50, and while under him, I got so bad that I could not pass urine only in drops, as —— or ——, had not passed a bougie into the bladder at all, but only down to the stricture, which was all they could do. One of them told me I would forever remain impotent.

When I had an emission at night no semen could pass. I always paid strict attention to all of their directions, as I had postponed my wedding-day three separate times. I then called on Dr. Larmont, 42 Reade st., who soon passed a bougie into the bladder without pain, and found three permanent indurated strictures. In two months from the time he commenced his treatment, I got married; my procreative organs became natural, and my wife enciente.

I hereby say, and am willing to take my oath, that I fully believe no other surgeon or physician could have cured me but Dr. Larmont, or, at any rate, without cutting the urethra open for at least five inches, the strictured portion of the penis; as I had consulted four or five other doctors who gave me that opinion.

J. H. A

New York, Dec. 3d, 1851.

This gentleman was presented, in due course of time, with one of the finest of boys, by his excellent lady, both of whom are in good health.

CASE XCIII.

Dr. Larmont cured me without pain, of syphilis, of some time standing, after I had used mercury for ten days previous, and all the time getting worse, yet he cured me in two days.

A. Waring.

New York, Nov. 5th, 1851.

CASE XCIV

Doctor Larmont cured me in a short time of gleet and strictures, without pain, which had existed for a number of months, notwithstanding my having paid hundreds of dollars, for the best medical services in the country, all to no purpose; for an affection or ulceration of my kidneys, which is the most harassing of diseases, for it cannot be reached only by the action of medicines.

M—— G——.

New York, May 10th, 1852.

CASE XCV.

About eight months ago, I came to Dr. Larmont, with a chancre and bubo, which I had been treated for to no benefit; he cured the chancre in a few days, but the bubo he had to open. After he doctored me a few weeks for it, and as it was not entirely well, and I had not much to do, I went to the hospital, and staid there two months, but still they could not cure me. I then returned to Dr. Larmont, who cured me in a short time.

H. Chabot.

New York, May 3d, 1852.

CASE XCVI.

A year ago, Dr. Larmont cured me in two days of chancres (syphilis) after I had been under other physicians for three weeks, but all the time getting worse He has now done the same again within that short time, notwithstanding I called on Dr. ——, corner C—— st. and Broadway; and was under his care for about three weeks, and every day getting worse.

George B——

New York, April 15th, 1852.

The above is another proof that quack remedies are impositions.

CASE XCVII.

Dr. Larmont:

I had been affected with an ulcerated leg, for some three years or more, arising from an old syphilitic disease, in connection with the abusive use of mercury, which laid me on my bed for a number of weeks, as it covered my entire leg from the knee down to the foot. Previous to the leg being so affected, I had ulcers on my body, head and face, for which I was constantly under

PLATE 27.

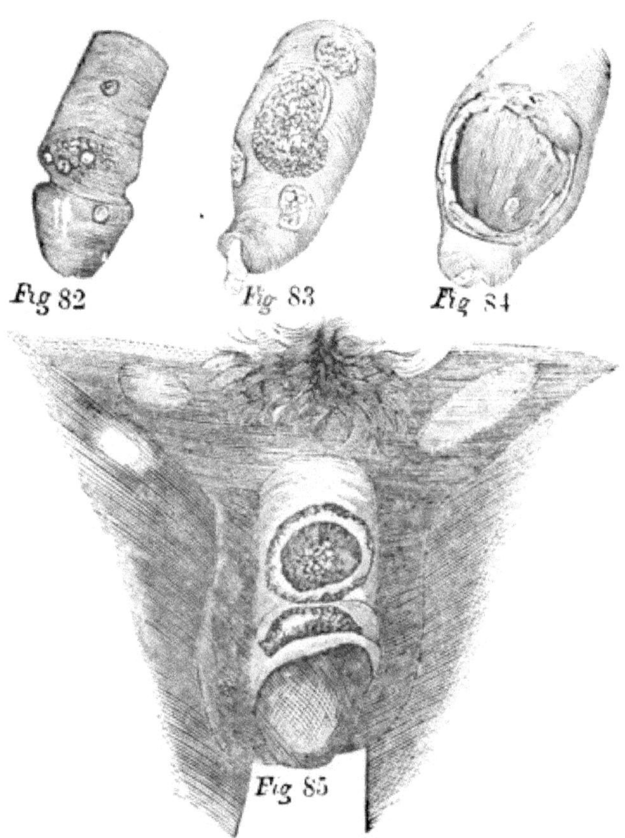

treatment by other physicians, to no benefit, as no sooner than one place was healed, another showed itself. In less than two weeks after commencing your treatment, I was able to go out, and attend to my business, and have been getting better, till I was well in body and mind in a few weeks. J. G. T——.

New York, Dec. 9th, 1851.

CASE XCVIII.

I contracted a gonorrhœa about six months ago, and applied to a number of physicians, besides using the quack remedies, but still the disease run into a gleet, which Dr. Larmont cured with one local application, yet I, at the same time, had a stricture, which he also cured. WM. K.

New York, April 24th, 1852.

CASE XCIX.

For a long time I was under Dr. M—— for the cure of gonorrhœa, and kept getting worse. When I first went to him he said he would cure me in a few days, as the disease had only just developed itself. I was unable to work for some time, and when I came to Dr. Larmont the inflammation of the bladder was so high as to cause me to pass water thirty times a day. Dr. Larmont cured me in a day, so that I went to work, and am now as well as ever. GEORGE D——.

New York, April 11th, 1850.

CASE C.

M. Larmont, surgeon, 42 Reade street, New York, cured me with one application, without pain, in a few hours, of an old gonorrhœa. I had been under the charge of two other physicians previous to applying to him, and used every one of the advertised remedies, yet all the time getting no better. A friend that Dr. L. had cured, sent me to him. GEORGE S——

New York, January 17th, 1851.

CASE CI.

Dr. Larmont cured me of an old syphilitic disease, without mercury, in a short time, after I had been under other physicians without benefit.

J. AUHCENBALT.

New York, November 14th, 1851.

CASE CII.

About a year ago Dr. Larmont cured me of constitutional syphilis, complicated with scrofula, without my being obliged to relinquish my daily occupation. Before applying to him I had been under the charge of Drs. L——, M—— and B——, for a long time, without any improvement. I applied to each in the order here named. It is now a number of years since I contracted the infection. W. V. L——.

New York, April 20th, 1852.

CASE CIII.

Dr. Larmont cured me in a few days of syphilis, after I had been under three other physicians for a long time, besides Dr. H——, and still getting worse. J. BURNS.

New York, January 20th, 1852.

CASE CIV.

For three years I had been afflicted with a gleet, three strictures so bad, that at one time I could not pass urine for twelve hours, and chancres, succeeded by constitutional syphilis. All the time I was under treatment in this and other cities, which cost me hundreds of dollars, and great loss of time, till I called on Dr. Larmont, who in a few days cured me of the chancres, and in a few weeks of the syphilis, sore throat, and other soreness. My throat was well in a few days, and the pains left me after I had taken but three pills. The gonorrhœa he brought away in one day, by one application, and the strictures he cured entirely in a few days. JAMES M——.

New York, May 1st, 1852.

CASE CV.

Dr. Larmont cured me of a gonorrhœa and stricture, of about six months' standing, in a short time. I had been under the charge of a number of doctors, without any benefit. JOHN T——

New York, January 7th, 1852.

CASE CVI.

Dr. Larmont cured me of an old gonorrhœa and stricture in a short time, after Dr. M—— tried to no purpose for three months. R—— S——.

New York, October 31st, 1851.

CASE CVII.

From a badly-treated gonorrhœa, the discharge continued for a long time, finally terminating in a gleet and stricture, with a weakness of the canal. After being under Drs. R—— and D—— L., as well as a number of physicians in other cities, Dr. Larmont cured me in a short time by local treatment, without using but very little medicine to strengthen the canal. His treatment was without caustic or pain. B——

New York, July 14th, 1851.

Charleston, S. C., is Mr. B's residence.

CASE CVIII.

Dr. Larmont cured me of an old gonorrhœa in a few hours, by one application, without pain. I had been under treatment before going to him. It is now some weeks since the cure. G—— S——.

New York, December 10th, 1851.

CASE CIX.

Dr. Larmont cured me of an old gonorrhœa, with one application, without pain, in a few hours after I had doctored for some time, as I had it for over six months, before coming to him. R. W. W——.

New York, Nov. 18th, 1851.

CASE CX.

Tioga Co., Feb., 1852.

Dr. Larmont:

Sir,—The bubo is well.

Respectfully, yours,

H——.

CASE CXI.

N——, Mass., Feb., 1852

M. Larmont:

There is no discharge for some time, and think I am well.

Yours, respectfully,

F——.

CASE CXII.

Canada West, Dec., 1852.

Doctor,—The sores are all healed.

Yours, C——.

CASE CXIII.

Dr. Larmont cured me of an old gonorrhœa, with one local application, after I had taken medicines, &c., to no benefit.

R. Ironsides.

CASE CXIV.

New York, April 6th, 1850.

Dr. Larmont:

I suffered the most any one could with syphilis, for over four months. I applied to a physician who said he would cure me in a week, as I had the sore only a day; but, instead of being cured in that time of the simple chancre, he done me no good, and the disease progressed till I had two bubos, one he allowed to break, which pained me awfully, and left five large hard blue places. After being laid up four months, I applied to you from the advice of friends, and in three days I was able to go to work; and you cured me in a few days longer, so perfectly, that no trace of it can be seen. All who do not wish to be kept in bed by other physicians, as I was, should apply to you, and they are certain of a speedy cure. Yours, most obediently,

F. L——.

CASE CXV.

Dr. Larmont cured me of constitutional syphilis, ulcers all over my body, and pains and debility, to such an extent as not only to deprive me of work for a month, but of sleep, only at short intervals of the night. I was all the time from the first appearance of it, first under Dr. ——, L—— street, to no benefit, then to two Thompsonian physicians, then to Dr's F—— & C——, and, still getting worse, I then called on Dr. Larmont, in the situation stated in the first part of this certificate, with the addition of gonorrhœa and stricture, which they all in succession treated without benefit. In one week after Dr. Larmont took me in charge, I commenced work, and increased in strength and health rapidly, till I was entirely cured of the whole, in two months' time.

J—— H—— D——.

New York, July 1st, 1851.

CASE CXVI.

This is to certify, that after paying Dr. H—, of D—— street, for curing me of gonorrhœa in its first stage, which he promised to do in three or four days, that he kept me taking the worst kind of medicines for two months, and still I was getting worse. He refused to return me my money, after he had tried everything, mercury included, as he himself told me. I then went to another place, and he still made me worse. I then, through the advice of a friend, called on Dr. Larmont, with a bubo and the worst kind of gonorrhœa, which pained me a good deal. Dr. L. cured me of the bubo in a few hours, and of the gonorrhœa in a few days, without pain or hindrance. C ——.

New York, August 16th, 1850.

CASE CXVII.

Dr. Larmont:

You have cured me as you said you would, of syphilis, caught about a year and a half ago, and for the cure of which I was under other physicians a long time, at a very heavy expense. This you have done without mercury, or even requiring me to lay by from work a single day. N. E. M——

New-York, July 23d, 1850.

CASE CXVIII.

This is to certify that Dr. Larmont cured me in a few hours, after I had the gonorrhœa, since the first of May last, and had been given up by Doctors J—— and H——; the last gave me drops for about seven weeks.

Wm. B——.

New York, July 20th, 1850.

CASE CXIX.

This is to certify that you cured me of a gonorrhœa of the canal of the penis, and of the anus, in a few days, after I had been under the care of another physician, for over two months. Eugene C——

New York, Sept. 11th, 1850.

CASE CXX.

Last December I contracted a gonorrhœa; as soon as I saw it, I commenced taking the different advertised remedies, which I have used up to this time, or within a day or two, till I came to Dr. Larmont, by the advice of a friend, and who cured me without pain, by using one application. The medicine I took previously had only changed the disease into the chronic or gleety state.

George M. S——.

New York, April 19, 1851.
Nine days after being cured.

CASE CXXI.

Dr. Larmont cured me of a gonorrhœa that I had taken medicines for, for over two months, without benefit, by a single application.

L—— L——.

New York, April 26th, 1851

CASE CXXII.

Dr. Larmont:

You cured me of a gonorrhœa with one application, after I had been under another physician over two weeks, to no benefit. N. McM——.

New York, Feb. 19th, 1851.

CASE CXXIII.

Six months ago, I applied to Dr. M—— and another physician for the cure of a gonorrhœa. They gave me medicines and injections, till it ended in gleet, which Dr. Larmont cured me of, with a single vegetable application, so that I was as well as I ever was, in a few hours. John W——.

New York, Sept. 30th, 1850.

CASE CXXIV

Dr. Larmont:

You cured me with one local application in a few hours, of a gonorrhœa. J. V. F——.

New York, Feb. 1st, 1851.

CASES CXXV and CXXVI.

Dr. Larmont:

You cured me in a few hours of gonorrhœa, by your local application, after I had been under another doctor for some time. You also cured a female in a dav or two, that had it. Welean Smith.

New York, Dec. 17th, 1850.

CASES CXXVII and CXXVIII.

Dr. Larmont:

Dear Sir,—You cured me of a gonorrhœa of over a year's standing, after another physician could not—by your single local application. In two other cases you did the same by one application. D—— S——.

Jersey City, Dec. 27th, 1850.

CASE CXXIX.

N. H., Conn., March 29th, 1851.

DEAR SIR :

About five weeks since I called on you, and got some remedies, which cured the female in one week, to the great joy and satisfaction of the person, who had taken medicine for three months to no purpose.

W— G.

To Dr. Larmont.

CASE CXXX.

Albany, June 17th, 1851.

DEAR SIR :

The remedy you gave me cured my complaint.

Yours, in haste,

S. R—.

To Dr. Larmont.

CASE CXXXI.

DR. LARMONT :

Dear Sir,—I am happy to say you cured me of gonorrhœa in a day or two, with one application without pain, after I had used the advertised remedies without avail. I should here say, I had a stricture for three years, which you also cured without caustic or pain, within two weeks. Through carelessness, in rubbing my eyes, I inoculated them with the gonorrhœa, which I am thankful to you for curing in two days.

J. C.—.

New York January 16th, 1850.

CASE CXXXII.

DR. LARMONT :

Dear Sir,—After having been under treatment for about three weeks without being cured of a gonorrhœa, I called on you, and am happy to state, that you stopped the discharge in five or six hours.

C. LAUDY.

New York, April 26th, 1850.

PLATE 28.

Fig. 86

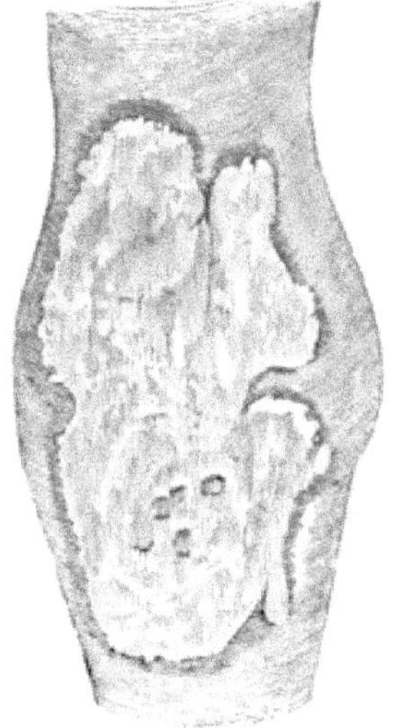

Fig. 87

Fig. 88

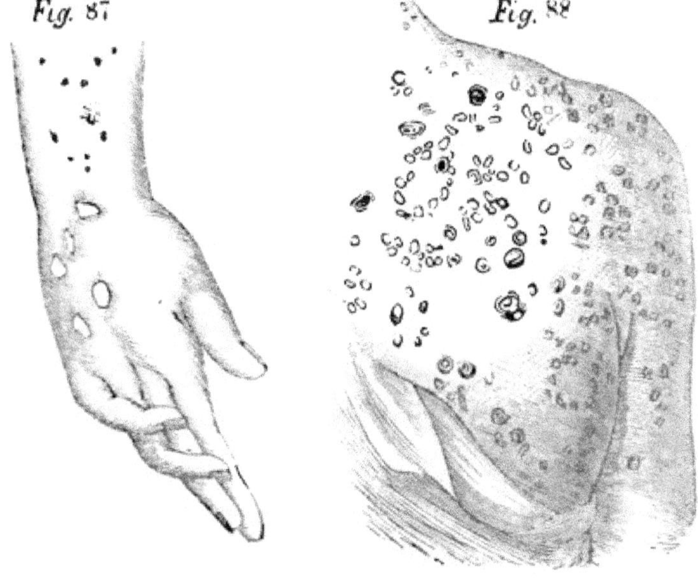

CASE CXXXIII.

Last June I was affected with a venereal disease, and applied to a physician for its cure. After being under his care for about nine months, I became so much worse, that one testicle was three times the usual size, and very painful. I then applied to Dr. Larmont, 42 Reade st., who entirely cured me in a very short time. N. F. E——, Brooklyn.

New York, May 16th, 1850.

CASE CXXXIV.

This is to certify, that I applied to Dr. H——, D—— street, for the cure of a delicate disease in its first stage He said he could cure me in a few days. I was under him for a long time without being cured. I then went to Dr. J——, D—— street, with the same result. From him I went to Dr. M——, W—— street, with no better success. The time I was under the treatment of the above persons was from February to October of 1849. I then called on Dr. M——, F—— street, Dr. R——, G—— street, and Dr. B——, Broadway, and they told me I could not be cured without an operation with the knife. I then called on Dr. Larmont, 42 Reade street, who told me he could cure me with medicines and local applications, without mercury, which he has done.

J. H. Demarest.

New York, Feb. 23d, 1850.

My space and time will not allow me to add but a few more names from the many certificates not mentioned in the previous pages, instead of copying them in full, for I have already given so many of all the diseases of a private nature, that all must be convinced of the unapproachable and unparalleled cures ever performed in this or any other country. I shall only give this additional one in detail.

CASE CXXXV.

For nearly four years, I had been afflicted with a general derangement of the whole system, such as indigestion, costiveness, impaired memory, great suffering and depression of spirits, a threatening destruction of the virile power, or with impotency, swelling of the testicles, seminal emissions, and a very frequent suffering and necessity of passing urine, both day and night, causing loss of sleep, and with a stricture of the urethra. During the above time I was under the charge, and followed the advice of a number of physicians, wearing trusses, bandages, using cold baths, yet constantly getting worse, till a friend sent me to Dr. Larmont, 42 Reade street, who, I am happy to say, has cured me in less than a month, by local treatment only. By the third night after his treatment commenced, I slept without getting up to urinate. I have been as well since the cure as I ever was. I can recommend full reliance on his medical and professional assurances. The bandages were all immediately laid off, on coming to him. C. C. B.

New York, June 10th, 1851.

CASES CXXXVI AND CXXXVII.

Mr. W. Fink and Mr. D. Harpen would be pleased to verify to any one, in person, the truth, that Dr. Larmont cured them by one vegetable application, in a few hours, of very bad gonorrhœa, which had been under treatment a long time previously.

I must acknowledge it is grátifying to receive such an evidence of proof, as the short extract below—and the letters in the foregoing pages—from brother members of the profession. It is the highest proof that can be rendered.

Rio Grande City, Texas, 22d July, 1852.

SIR:

I am informed that you have a preparation of great use, in the cure of venereal affections, particularly stricture and genital debility. Large quantities can be disposed of, on this frontier. Our balsamic preparations, do not fill the desired purpose, of a certain and speedy cure of gonorrhœa. I am a practicing physician, a graduate of the Jefferson Medical College, Philadelphia, and would be pleased to have the use of such remedies, if you are disposed to establish an agency in this country. Your directions in relation to terms, use, &c., will be in all cases strictly followed out.

Very respectfully,

THOS. P. LINTON, M. D.

DR. M LARMONT, New York.

CHAPTER XXIV.

A FFW WORDS OF ADVICE TO PATIENTS AFFECTED WITH WHAT ARE GENERALLY TERMED "VENEREAL DISEASES."

I do not think it is necessary for me to present any further remarks in reference to my own ability and superior manner of curing all of these poisonous affections —in an incredible (almost) short space of time—than the unmistakable proofs presented by the large number of certificates—covering every private infection or disease of the genital organs—acknowledgments from physicians, and the editorial testimonials which will complete the work. These notices were published without my knowledge, as the notices themselves say, they are representations of some of the most eminent physicians who are acquainted with my practice.

I wish to put every afflicted person on their guard against designing impostors and the quack or advertised remedies which nearly fill a column of some of the city papers, not only lauding their pretended virtues, but with a baseness that is really surprising, assert that they are the only known cure in the world. Such a declaration alone should convince any one that they are worthless and dangerous compounds, which my numerous certificates so truthfully verify. In conclusion, if any one, after reading this work, still places confidence in quack remedies, they can go to any drug store and get them made up for themselves at the trifling expense, probably, of twenty-five cents

A SAMPLE OF THE INGREDIENTS, COMPOSING THE DIFFERENT ADVERTISED OR QUACK REMEDIES.

Red Drops.—This is a dangerous preparation, consisting of corrosive sublimate, dissolved in alcohol, and colored with compound spirits of lavender.

Unfortunate's Friend.—Make a decoction of sarsaparilla root, rasped guaicum, elder flowers and resins, of each a quarter of a pound; mazerian root, burdock root, parsley root, and whortleberry leaves, of each two ounces; liquorice root, half a pound. Put them all together in one gallon of water, and boil in a covered vessel down to a quart, and strain; while the decoction is warm add of balsam copaiva, sweet spirits of nitre, of each half a pound; powdered gum Arabic, white sugar cubebs and carbonate of soda, of each a quarter of a pound. Dose—a wine-glass full occasionally.

Every one will see by the above that all quack remedies are composed of either mercury, copaiva or cubebs, disguised by mixing them with other remedies. I have a hundred different prescriptions which I could give if time and space allowed, every one of which contains either the one or the other of these three articles, which some combine with injections and washes, composed of mercury, caustics, alum, lead, zinc, &c., under the head of lotions. As another instance of the deception practiced in this and other cities in private practice, I will mention the fact of one man having three offices, under as many different names.

With three or four exceptions, there are no regularly educated physicians in this practice, besides myself, in the city, notwithstanding the great number of names in the papers assuming to be such.

Ricord of Paris, Acton of London, and every European surgeon and physician of standing who have published works on private diseases, say there are five varieties of primary syphilis, and two when they have become constitutional; and that each kind requires a different treatment. How persons, then, who have not had the necessary experience in the cure of these diseases, expect to be able to rid their patients of such a dangerous affection, every sensible unfortunate had better find out before they put themselves under their care.

Gonorrhœa, blennorrhagia or clap, is as distinct a disease from syphilis, and requires as distinct treatment as small-pox and cholera. Syphilis affects the system, but the clap is local in its action, producing such local complications as swelling of the testicles, phimosis, paraphimosis, inflammation of the bladder, prostate gland, stricture, bubo and gleet. The eyes may sometimes be affected sympathetically.

Each one of these stages or complaints require different treatment. All the medicine in the world cannot cure a stricture, or gleet. The same medicines that might be beneficial in a case of clap has no effect on a gleet, or the nose or eyes when they have been inoculated with the discharge. The eyes have been destroyed in twenty-four hours by the rapid progress of the disease, when direct inoculation has taken place. Any person, then, who is acquainted with these diseases, and sells a quack remedy as a cure-all, do so to get your money—and without the remotest idea that you will be benefited by taking them.

I have not given full descriptions of the various forms of venereal diseases; nor have I referred to my treatment, or entered into the details of the various modes of relief. As I have often said, each case requires that treatment which the symptoms and disease indicate; and medicines in the hands of the ignorant, irresolute, or timid patient, does more harm than good. In this way only can we hope to rescue the afflicted from making a degrading application to a more degraded quack, or the unfeeling and dangerous routine of the regular but inexperienced graduate.

DON'T FAIL TO READ THIS ADVICE TO THE AFFLICTED

The moment you discover that you have contracted a private disease, or if you have had any affection of the kind, at any time previous—even years before—and which you have supposed you were cured of, by your country or other physicians, apply to me for this reason. Few physicians have ever been taught anything about the treatment of venereal diseases. Even if they had, it was that of the old mercurial or copavia remedies, and which often causes more injury and suffering than the original disease. Further—this, as well as more enlarged works, too plainly show, that many are pronounced cured—by inexperienced physicians—who get married, the disease is reproduced by the time which may have elapsed, and the extra excitement such an event generally produces, and the unsuspecting victim finds that he is yet affected, also the child, if the wife happens to be pregnant. In some cases, the child may not show any signs of being affected for some years after it has been born. Sooner or later, however, it will show itself in the whole circle, if the original complaint was not entirely eradicated. Or if you have had an emission involuntarily. Sit down and write me a full statement, by giving your age and sex—single or married; when you had the suspicious connection, and when you cohabited with your wife last. Whether bilious or nervous temperament; complexion, habits, and occupation. Then state the case, symptoms, duration of illness, and supposed cause, and whether your bowels are regular. Then refrain from everything that is stimulating; keep the parts clean, and be careful not to inoculate the eyes, nose, anus, or any other part, with the poison.

I can then send you the necessary remedies by mail or express—state which you prefer—in time to check, and permanently cure you at once, even if you are in the remotest part of the Union or British Provinces.

All my packages sent are sealed, so as to be proof against detection; and as they are so rapid and convenient in destroying the disease, you can cure yourself, even amongst the most fastidious friends, with perfect secrecy

Letters for advice or treatment must contain five dollars, or they cannot be noticed, as I am so much occupied that I cannot even read letters that do not pay me for my time. Those who reside in small or inquisitive places, need only write the following on their letters: "Box 844, New York Post Office." I shall get them just as safe as I would if my own name was superscribed on them.

Patients who apply personally, should be careful and notice my office.

N. B.—My office is divided off into separate rooms, so that patients are only seen by myself.

If I had supposed it would have been necessary, or interesting to those who may find this unpretending work sufficiently instructive to read its entire contents, every letter and illustrative case included, I would have continued the letters and cases to as many more. The reader or patient will please understand, however, that I do not preserve any letters, without the patients or my correspondents desire me to publish them as additional proofs of my superior practice, for the purpose of convincing those who may have given up all hope of ever finding a physician who could cure them. I have letters almost daily from such unfortunates who have been under the care of different physicians for years and years, all this time, however, loosing strength and health. If such persons could call in person, I could present to them living proofs, as well as the written ones, of the invariable success of my treatment. I will therefore only say in conclusion, that I can offer you such treatment, as it appears cannot be obtained anywhere else in America.

But a very few physicians correctly understand the different complaints, the delicate constitution of females are subject to. The most frequent complaints to which they are liable, are the irregularity of the menses, and prolapsus uteri, or a falling of the womb, accompanied or produced by fluor albus, or whites. Either of these difficulties will cause consumption, if they are neglected, or allowed to progress, and the proper treatment for their cure not understood.

Those who wish to be cured of fluor albus, (whites,) or prolapsus uteri, (falling of the womb,) can do so by giving me a full statement of their case.

$5.00 consultation fee must be remitted when making the application, or the value of my time will prevent its being noticed.

CERTIFICATES.

From a German Physician.

NEW YORK, December 12, 1853.

DR. LARMONT.

Dear Sir,—I know all about the best treatment of the diseases upon which your work treats, as I resided in France several years, but still know nothing equal to the cures effected by you.

Respectfully,

DR. J. D.

From a Physician in Florida.

I treated myself a number of months for the cure of a Chronic Gonorrhœa without success, and hearing from patient of mine that Dr. Larmont had cured him and others of his acquaintances in a very short time, I resolved—as I was to be married in three weeks—not to lose any more time, but to extend my visit as far north as New York, and put myself under his care at once. It is with the greatest pleasure that I bear witness to his having cured me entirely, in forty-eight hours, and of course advise all who are similarly situated to em ploy him, to render them the same kind services.

DR. A. G.

From an old Physician in Virginia.

It is with heartfelt gratitude that I state an additional remarkable cure of Dr. Larmont, as he cured me in less than two months of Impotency, with all its complicated general derangements, which I had labored under for many years, notwithstanding all the treatment which my position, as an old physician, enabled me to obtain. It is about a year since the cure, and I therefore know it to be permanent, though my age is very far advanced.

DR. C—— C.

I can only add short extracts from two or three other Certificates of Cures.

I was treated by other doctors three months, for the cure of a Clap, but Dr. Larmont cured me in one day.

CASPER GEISINGER,
40 Delancy street.

I had been under the care of about thirty physicians within twelve years, to be cured of Impotency and its various complications, yet finally, I was obliged to go to Dr. Larmont, who cured me in about two months.

J. S. R.

I was totally Impotent for over 8 years, the semen was almost constantly oozing from the urethra; I had been under the care of a number of Professors in the Medical Colleges, and eminent physicians in this and other cities, and was given up as having Consumption, (having a cough,) Heart disease, and almost every symptom of disease that the human system is affected with, but Dr. Larmont cured me entirely in about three months without cauterization.

F. S.

My brother and myself have used the humbug instrument, which a doctor advertises will cure seminal diseases, but I grew worse, and he became insane from the seminal loss. We used them nearly two years.

G. A. Y., Connecticut.

Dr. Larmont cured me of Diurnal and Nocturnal emissions in one week, after I had been given up as incurable.

J. G.

I was cured by Dr. Larmont, of Impotency and Diurnal Emissions in one week, by local treatment only.

J. E. P., Vermont.

Constitutional Syphilis caused my body to be covered with deep Ulcers. I paid the most eminent professor, physician and surgeon in this country $100, and was under his charge one year to no benefit, as my body was the same when I went to Dr. Larmont, but he cured me in about three months.

G. L., New Jersey.

We advise every person—whether married or single—to procure a copy of the Medical Adviser and Marriage Guide, illustrated with colored anatomical plates, and numerous cases. The author, Mr. Larmont, Physician and Surgeon, mails them free to any address on the receipt of $1.

Opinions of the Press editorially, and many eminent Physicians, in favor of my superior ability.

From the Rahway Advocate and Register.

We wish to call the attention of our readers to the advertisement of M. Larmont, Surgeon, of 42 Reade street, New York. We have seen the manuscript of his forthcoming work, and are satisfied that a knowledge of its contents will be the means of rescuing thousands of individuals from an early grave. This is also the opinion of a number of eminent New York physicians that we have met, who are acquainted with his new and superior mode of treating all diseases of a private nature, for it is time that quackery should be suppressed. Dr. L's reputation for quick cures is well-established, not only in the city, but in all the States, the West Indies, the British Possessions, etc.

From the Essex (Elizabethtown,) Standard.

We are assured by a friend, in whose judgment we place great confidence, that the doctor's forthcoming work will be found worthy of the attention of all those for whose benefit it is designed.

From the New Brunswick Times

We have known Dr. Larmont for a number of years, and not only coincide with the above, but are well convinced that the assertions made in his advertisement (in another column,) may be relied on as strictly true. Those interested, will, therefore, be able to shun the impostors who infest New York and other cities.

From the National Police Gazette, New York.

We take pleasure in referring to the advertisement of Dr. Larmont, in another column

From the Rahway Advocate and Register.

Dr. Larmont, of 42 Reade street, New York, has fully satisfied us that stricture of the urethra is one of the most dangerous complaints that man is afflicted with. The reason is, every one is liable, for it is not caused by private disease only, in more than a tithe of the cases; another and greater reason is, its progress and fatal development is so insidious that the victim is unconscious of danger, till he is at the brink of eternity.

From the Rahway Advocate and Register.

Diseases of a private nature, as almost every person is aware, are entirely different from those of a general character; that is, the private one is entirely local at first, but if allowed to remain uncured for more than a very few days indeed, is absorbed into the system, and of course becomes general. It will be seen by Dr. Larmont's new advertisement, that he publishes some remarkable certificates of immediate cures.

From the Rahway Advocate and Register.

We are satisfied that there are many innocent, or, in other words, diseases of the generative system, which all are liable to have however moral a life they may lead, and we are further satisfied that a physician like Dr. Larmont, of 42 Reade street, New-York, who has received his education from the highest sources, and who has for years devoted his entire attention to their cure, is fully able to treat them successfully.

From the Essex (Elizabethtown) Standard, N. J.

We fully agree with the above editorial notice, and can say further, that Dr. Larmont's advertisement, in another column, says no more than he does daily.

From the New York Day Book.

An eminent physician of this city, who has been acquainted with Dr. Larmont, of 42 Reade street, for a number of years, assures us that his treatment of those diseases, belonging to that specially arising from indiscretion, &c., is unequaled. We, therefore, fully indorse the encomiums of the New Brunswick Times, Essex Elizabethtown Standard, and Rahway Advocate and Register, of N. J.

We take pleasure in recording merit, and especially so when acknowledged oy those of the same profession, and for that reason copy the above in reference to the ability of Dr. Larmont, 42 Reade street.—*Notice from the Sunday Dispatch.*

From the National Democrat.

Being convinced, from the highest professional as well as other assurances, the encomiums of the New York Day Book, Essex Standard, New Brunswick Times, Rahway Advocate and Register, (of New Jersey,) and the Sunday Dispatch, are only an acknowledgment of superior merit, we fully acquiesce in them in regard to Dr. Larmont of 42 Reade street, New York.

Editorial Extract from the New York Staatszeitung.

Dr. Larmont's system of curing Venereal Diseases is the quickest, cheapest, and surest of any with which we are acquainted. Dr. Larmont receives patients of either sex at his Office, 42 Reade street, corner Broadway. The latest cures that have been effected by the Doctor have established his reputation on so firm a basis, that his time is fully occupied by the most respectable patients.

From the New York Courier Des Etats Unis, July 23, 1853.

We suppose to do some service to the public in calling their attention to the work advertised to-day in our columns under the title of "Medical Adviser and Marriage Guide." Its author, Dr. Larmont, is a surgeon whose name is already known throughout Europe, and the work is written with as much conscience as science. The Doctor does not think he has done enough yet to humanity with his pen, for he can be consulted every day for all kinds of diseases.

Editorial Notice of the Reform (German Daily.)

Dr. Larmont's method of cureing all Venereal Diseases,—as we are assured by many persons who have had the advantage of his professional abilities—is the quickest and surest of all. We beg to call the attention of our readers to his advertisement in another column of our paper.

From the Empire City.

We warned our readers against the Quack impostors, alias Doctors, who post up their decoy obscenities on the lamp posts, fences, etc., in violation of the city ordinances; and the better to prevent the afflicted from falling into their snares, advise them to read Dr. Larmont's advertisement in another column, and purchase his valuable work, as we coincide fully with the Editorial's of all the papers, referred to in his work.

THE MEDICAL ADVISER.—This book, as published by Dr. Larmont, purports to give a synopsis of the causes, symptoms, and most certain cure of all those diseases brought on by the indiscretions of youth and to which they are especially subjected in a city like this. The price is only one dollar.—*National Police Gazette.*

DOCTOR LARMONT'S WORK, advertised in another column should be in the hands of every person old enough to read. We know there are hundreds of thousands who are bringing disease upon themselves without knowing it, as family or general practitioners have not the opportunity to investigate the cause of the speciality of which this work treats.—*National Police Gazette.*

* * We not only agree to all of the above, but take pleasure in saying that Surgeon Larmont has sold thousands of his valuable work within a few months, in the different sections of this great country. He mails his Medical Adviser and Marriage Guide, with colored anatomical plates, which every person should read, to any address free of postage, on the receipt of one dollar Send for it.—*N. Y. Pick.*

Testimonials from physicians of the most eminent ability can be seen, with his Diploma, at his office. No one now, after all these evidences, need go astray.

HIGHLY IMPORTANT
TO THE MARRIED
AND THOSE CONTEMPLATING MARRIAGE.

From the *New York Atlas*, of June 15, 1856.

PARIS AND LONDON MEDICAL ADVISER AND MARRIAGE GUIDE. 366 pages. Twentieth edition. 12mo., cloth. Profusely illustrated with nearly 100 beautiful electrotype engravings. By M. Larmont, Physician and Surgeon, New York.

Medical literature has been greatly depressed by the frequent periodical avalanches of medical books and pamphlets, which have buried the truth in antediluvian darkness. The false teachings of these works have misguided the public, depraved the taste and refinement of the young, and so distorted nature's laws, as to cause them to be despised and trampled upon.

We therefore take the more pleasure, in a brief notice of the Medical Adviser and Marriage Guide, in commending it upon its real merits. It holds the mirror up to nature, in good earnest, anatomically and physiologically.

To the young as well as to the married, it is replete with interest and information.

The particular department treated of is that of the Genito-Urinary organs and their diseases, which are ably treated by the doctor.

To the married and those who contemplate forming those holy bonds, the work is of the most inestimable value. Price $1, and four letter stamps, to pre-pay postage.

SURGERY.—We call attention to the notice of a work, on our outside, by Dr. Larmont, an experienced surgeon of this city. In curing the diseases to which the doctor pays his particular attention, he has no superior in this country, if he has an equal.

The following unsolicited editorial is from the *National Police Gazette* of January 16, 1857:

MEDICAL ADVISER AND MARRIAGE GUIDE: by M. Larmont, corner of Spring and Mercer streets, New York City. This work is an awful warning to those who run the risk of contaminating themselves by breaking the laws of nature and society. It presents the whole anatomy of man in his relations to his family and offspring, and, seriously considered, such a work if widely circulated might save the blood of a large portion of the race from corruption and deterioration. Price one dollar.

And was copied by the *New-Yorker Criminal-Zeitung and Belletristisches Journal* of February 4, 1858, as an editorial. And by the *New York Staats-Zeitung* of February 11, 1858, editori ally.

PARTICULAR NOTICE.

In consequence of recent complaints of patients who have lost money letters, and other letters not having been answered by us, I found, on inquiry, that there is an *itinerant Doctor* traveling about the country, and in this city, by the name of Larmont; therefore I wish all letters to be directed to my associate (who has been with me for several years), Dr. E. Banister, box 844, P. O., New York City, as formerly

M. LARMONT,
Physician and Surgeon.

N. B.—Patients are respectfully informed that our offices have been removed from 82 Mercer Street, corner of Spring, to 647 Broadway, *up stairs* (La Farge House block), one block above the Metropolitan, and *three* blocks above the St. Nicholas Hotel, New York City.

CONTENTS.

PART I.

THE ANATOMY OF THE GENERATIVE ORGANS OF THE MALE AND FEMALE.

MALE.

PAGE.